D0398681

System Enquiry

System Enquiry

A System Dynamics Approach

Eric F. Wolstenholme

Foreword by Jay W. Forrester

JOHN WILEY & SONS

Chichester · New York · Brisbane · Toronto · Singapore

Other Wiley Editorial Offices

John Wiley & Sons, Inc., 605 Third Avenue,
New York, NY 10158-0012, USA

Jacaranda Wiley Ltd, G.P.O. Box 859, Brisbane,
Queensland 4001, Australia

John Wiley & Sons (Canada) Ltd, 22 Worcester Road,
Rexdale, Ontario M9W 1L1, Canada

John Wiley & Sons (SEA) Pte Ltd, 37 Jalan Pemimpin #05-04,
Block B, Union Industrial Building, Singapore 2057

Library of Congress Cataloging-in-Publication Data:

Wolstenholme, Eric F.
    System enquiry : a system dynamics approach / Eric F. Wolstenholme
; foreword by Jay W. Forrester.
        p.   cm.
    Includes index.
    Includes bibliographical references.
    ISBN 0-471-92783-X
    1. Systems engineering.   2. System analysis.   I. Title.
TA168.W647   1990
003—dc20   90–35828
                                                               CIP

British Library Cataloguing in Publication Data:

Wolstenholme, Eric F.
    System enquiry : a system dynamics approach.
    1. Systems. Dynamics. Mathematical models
    I. Title
    003

    ISBN 0-471-92783-X

Printed and bound in Great Britain by Biddles Ltd, Guildford, Surrey

To

Liz, Mat, Tom and Susie

Contents

Foreword

Our social, economic and industrial systems are growing ever more complex. The baffling nature of such systems lies behind the troublesome symptoms reported in the daily press-corporate failures, urban decay, defaults on debt, terrorism, drug addiction and environmental degradation.

Since the pioneering work on System Dynamics at MIT beginning in 1956, the System Dynamics field has been devoted to developing methods for better understanding of our physical and social systems. New developments in the field are now coming from many parts of the world. More than thirty books have appeared ranging from theory through applications, even so, System Dynamics is still in the early stages of development. Teaching materials in the form of texts, readings, lecture aids, and video still fall short of the need.

Many leaders in the field are beginning to develop materials to help people to see how their policies are often producing the opposite of the desired results. Some are leading exciting experiments to find ways that System Dynamics can be made a core framework for teaching in the primary and secondary schools by providing a common foundation to unify mathematics, physical science, social studies, historical change and a student's personal life experiences. Others, for the corporate audience, are developing computerised simulations that take the next major step beyond case studies as a way to learn about management through concentrated vicarious experience.

In this book, Eric F. Wolstenholme addresses another gap in the literature of systems. He sets System Dynamics in a broader framework that includes operations research and the descriptive treatment of social and economic systems. This is long overdue. Unnecessary and counterproductive barriers have grown up between disciplines that are all, from different starting points, striving toward the same ends. By providing bridges between the several approaches to system behaviour, Wolstenholme helps those in different disciplines to join forces toward an improved social and economic future.

Eric Wolstenholme has made major contributions to System Dynamics during the last decade. In addition to his technical contributions as a manager, consultant and modeller, he is a founder of the System Dynamics Society in which he has played an influential role. Following his experience in producing and editing the world's first System Dynamics journal, 'DYNAMICA' at the University of Bradford, he became Executive Editor of the Society's journal, *System Dynamics Review* during its inaugural three years. During 1989 he was President of the Society. The field of System Dynamics owes much to Eric Wolstenholme for his leadership and initiative.

Jay W. Forrester
October 1989

Preface

I have a number of strong motivations for writing this book.

The prime reason arises from my experiences in using three different approaches to modelling problem situations. These are Operational Research, Soft Systems Problem Solving Methodologies and System Dynamics. Encounters with these subjects has led me to a sound belief that System Dynamics has a much more extensive contribution to make to the field of System Enquiry than previously perceived or achieved.

It is interesting to view System Dynamics from the perspective of the two other subjects referred to. Due to its origins in control engineering and computers it is primarily seen by the Soft Systems field as being firmly aligned with the Hard Systems or science based field of enquiry of which Operational Research forms a large part. However, as a result of its ability to incorporate subjective elements in models and its adoption of a wide definition of model validity, it has created for itself an image in the eyes of hard system modellers of being a soft system type of tool.

It is my opinion that it can fulfil a role as both a soft and a hard approach to System Enquiry. Consequently this book formalises both aspects of the subject and presents both Qualitative System Dynamics and Quantitative System Dynamics. The former is centred on diagrammatic modelling as a means of describing and analysing complex systems and is described in Chapters 2 and 3. The latter is based on converting these diagrams into formal simulation models. The techniques of simulation are described in Chapters 4 and 5 and case studies are presented in Chapters 6, 7 and 8 using both the DYSMAP2 and STELLA simulation languages.

Secondly, I have a desire to take the mystery out of where System Dynamics models came from. The model conceptualisation process is by far the most difficult task in the application of the System Dynamics method, consequently an attempt is made in Chapter 3 to improve the communication of this process.

The third motivation in writing the book was to try to explain some of the many insights which are possible from employing the System Dynamics method. Chapter 3 focuses on using Qualitative System Dynamics for policy design and for anticipating 'counter intuitive behaviour' of systems.

Chapter 7 concentrates on using Quantitative System Dynamics for policy design. In particular, it shows how System Dynamics can be used to quantify savings in a physical capacity arising from improved control and to identify and use isomorphic structures. Chapters 6 and 8 show how simulation models can be used to improve understanding by quantifying the effects of alternative system operating policies.

Finally, I have a wish to present the use of optimisation analysis as a method of policy design in System Dynamics models. This has been one of my most significant areas of research over the past few years. Chapter 9 introduces the concepts and Chapter 10 presents an extension to the case study of Chapter 8, to demonstrate the additional insights which can be achieved by optimisation over conventional System Dynamics policy analysis.

The book is intended for Policy Makers, Modelling Analysts and Students in Strategy and Policy Analysis and in Management Science.

Professional Policy Makers and Students in Strategy and Policy analysis should read Chapters 1, 2 and 3 and the case studies in Chapters 6, 7, 8 and 10. Analysts and Students of Management Science who are unfamiliar with System Dynamics should read all chapters plus the Appendices, which contain detail listings of the models developed throughout the book. Analysts and Students of Management Science conversant with System Dynamics should focus on the case studies of Chapters 6, 7, 8 and 10 and the Appendices.

The introduction to Quantitative System Dynamics in Chapters 4 and 5 is, purposely, presented at a basic level to encourage readers with limited quantitative ability to sample the benefits of this stage of model building. More advanced quantification and equation formulation is introduced during the later case studies.

The subject of System Dynamics is so generally applicable that it is difficult in a book of this size to select subject areas for applications which will be of interest to all readers. Examples of applications have been chosen from as wide a field as possible and it is hoped that this selection is sufficiently broad to demonstrate the breadth of application of the approach and to indicate the transferability of the ideas to other fields. The fields covered include Manufacturing, Research and Development, Social Services, Health, Defence and Mining.

The book assumes no previous knowledge of System Dynamics on the part of the reader and only a general appreciation of algebra is necessary to understand the simulation languages and simulation models used.

The essential prerequisites are an open mind and a commitment to improving the quality of thinking and action. The tools presented are simply a means to this end.

E. F. Wolstenholme
June 1990

Acknowledgements

First and foremost my thanks are due to Professor Jay W. Forrester, recently retired from the post of Germeshausen Professor at the Sloan School of Management, Massachusetts Institute of Technology, Boston, USA, for creating the conceptual framework and methodology which has come to be known as 'System Dynamics'. This has had a profound effect on my life. I am particularly grateful to Professor Forrester for providing the foreword to this book.

Secondly, my thanks are due to two ex-colleagues at Bradford University Management Centre for helping to develop the work on which the book is based. These are Professor R. G. Coyle, now Professor of Defence Strategic Analysis, Royal Military College, Shrivenham, UK, who introduced me to the subject of System Dynamics and Professor J. C. Higgins, recently retired from the Directorship of the Management Centre and the Chair of Management Science, who provided invaluable support for my activities.

With regards to the production of the book itself, I would like to thank a number of people for their contributions. The reviewers of the proposal of the book and my close colleagues in the System Dynamics world all made valuable and constructive suggestions. In particular, I am indebted to Dr J. D. W. Morecroft of the London Business School, UK and Dr B. C. Dangerfield of the University of Salford, UK for their copious comments on my original draft, and to Dr F. Wheeler and Mrs K. M. Watts of the University of Bradford Management Centre, UK for their helpful suggestions.

From a technical point of view I would like to express my gratitude to Professor R. Keloharju, of the Helsinki School of Economics, Finland. Not only for his general and much under-rated contribution to the field of optimisation in System Dynamics and the development of the DYSMOD software, but also for his collaboration with me in developing applications of this technique.

Additionally, I would like to acknowledge the contributions of Dr A. K. Ratnatunga and Dr J. Zachoval to Chapter 7 and Dr A. S. Al-Alusi to Chapter 10.

I am also grateful to my wife Liz Wolstenholme of Yorkshire Health for the many helpful professional discussions on the identification of feedback structures and for suggesting appropriate examples to use in this book I take full responsibility for the interpretations I have placed on these.

A major factor in the production of this book has been the work of my secretary Mrs K. Cousens. Without her patience in typing and word processing my deadlines would never have been met.

Finally, it is very clear to me that the book would never have been written without the private support of my family, to whom the book is dedicated.

The basic ideas underlying Qualitative System Dynamics in Chapters 2 and 3 were first published in a joint paper with Dr R. G. Coyle entitled 'The Development of System Dynamics as a Methodology for System Description and Qualitative Analysis' ((1983) *Journal of the Operational Research Society*, **34**, pp 569–581). Material from this source is used with the permission of Pergamon Books Limited. These ideas were subsequently developed in a paper entitled 'An Overview of System Dynamics' ((1989) *Transactions of the Institute of Measurement and Control*, **11**) and material from this source is published with the permission of this Institute.

Early versions of the Coal Clearance Model in Chapter 7 and the Armoured Advanced Model in Chapter 8, were first published in the *European Journal of the Operational Research Society* in papers entitled 'The Relevance of System Dynamics to Engineering System Design', ((1983), **14**, 116–126) and 'Defence Operational Analysis Using System Dynamics' ((1988) **34**), respectively. Material from these sources is used with the permission of Elsevier Science Publications.

The material in Chapter 7 concerning algorithmic control modules was first published in the *System Dynamics Review*, in a paper entitled 'Algorithmic Control Modules for System Dynamics Models', ((1986) **2**).

Background work on the optimisation of defence models as described in Chapter 10, was also first published in the *System Dynamics Review* in a paper entitled 'System Dynamics and Heuristic Optimisation in Defence' ((1987) **3**). Some material from this paper is used in Chapter 10 with the permission of the System Dynamics Society, but the copyright of the enhanced work presented here resides with the Controller HMSO, London, 1989.

The material in Chapter 9 on optimisation in System Dynamics was first published jointly with R. Keloharju in the journal *Systems Practice* in a paper entitled 'The Basic Concepts of System Dynamics Optimisation' ((1988) **1**, 65–82). This material is used with the permission of Plenum Publishing Corporation.

Appendix 6, contains various extracts from the DYSMAP2 User Manual and is published with the permission of the University of Salford, UK.

The software packages described in this book are copyrighted, with all rights reserved. DYSMAP and DYSMOD are copyrighted to, and are trade marks of, the University of Bradford, Richmond Road, Bradford, England. DYSMAP2 is copyrighted to the University of Salford, Computer Services Section, Salford, UK and was written by Olga Vapenikova. STELLA and STELLAstack are copyrighted to, and a trade mark of, High Performance Systems Inc., Lyme, New Hampshire 03768, USA. HyperCard is a trademark of Apple Computer, Inc.

Chapter 1

An Overview of System Dynamics

INTRODUCTION

The purpose of this book is to present an approach to the general field of System Enquiry.

System Enquiry is a term defined here to describe a rapidly emerging field, which focuses on problem solving and analysis of complex real world systems by methodological means, where the emphasis is on promoting holistic understanding rather than piecemeal solutions. The field is alternatively known as that of large scale system analysis.

Some definitions of the terminology are obviously appropriate. The word 'system' is used here to denote any combination of real world elements which together have a purpose and which form a set which is of interest to the inquirer. Enquiry means a careful and diligent search which will withstand scrutiny and methodology implies a stepwise procedure for investigation which is independent of the content of investigation. The word holistic refers to the importance of the whole of a system and the interdependence of its parts.

It will be apparent from these definitions that the field of System Enquiry is a rather idealistic one, since it requires a totally multidisciplinary and general approach to be applied. Whilst this is true, there are methods which approximate to its requirements and the one addressed in this book is known as System Dynamics.

The dynamics (or behaviour) of systems can be studied at many differing degrees of mathematical sophistication. The approach used in this book is centred on the ideas of Forrester (1961, 1968), who created a subject area originally known as Industrial Dynamics, but now referred to as System Dynamics. This creation was in response to a recognition that many problem solving methods, particularly those linked to Management Science, were not delivering their promise of providing insight and understanding into strategic problems in complex systems.

System Dynamics means exactly what its name implies. It is concerned with creating models or representations of real world systems of all kinds and studying their dynamics (or behaviour). In particular, it is concerned with improving (controlling) problematic system behaviour.

The method is aimed at providing a distinctive set of easily usable tools which might be used by system owners, rather than just analysts; centred on a very generic set of building blocks which are universally applicable.

The purpose in applying System Dynamics is to facilitate understanding of the relationship between the behaviour of a system over time and its underlying structure and strategies/policies/decision rules.

The words strategy, policy and decision rule will be used synonymously throughout this book. The difference between them is essentially the time horizon over which they apply. Strategy is used to specify long term action, to move towards an objective, whereas policy or decision rule is used to specify short term action. Hence, the choice between the words in any particular context will depend on the time frame of that context.

In summary, the procedure is to observe and identify problematic behaviour of a system over time and to create a valid diagrammatic representation (or model) of the system, capable of reproducing (by computer simulation) the existing system behaviour and of facilitating the design of improved system behaviour. For example, changing behaviour from decline to growth or from oscillations to stability.

During the early years of development of the method, applications were largely industrial (Forrester 1961). Later the subject broadened (Forrester 1968) and a number of global and other large scale studies emerged (Forrester 1969, 1971; Meadows *et al*. 1972). During the late 1970s and 1980s, the scale of individual studies has been reduced, but the scope of application of the method has become extremely wide, covering most traditional academic disciplines of study, but with a strong emphasis on socio-economic areas. (Coyle 1977, Richardson and Pugh 1981, Forrester *et al*. 1983, Lyneis 1980, Roberts 1978, Roberts *et al*. 1983). The subject now has its own international society, and journal (the *System Dynamics Review*) and links between System Dynamics and other fields are rapidly being forged.

In addition to a broadening of applications of the traditional method, there has emerged in recent years, a broadening of the method itself. In particular, there has been a move away from an obligatory use of quantified simulation models towards an increasing recognition of the relevance of the diagramming phase of the subject (Wolstenholme 1982, Wolstenholme and Coyle 1983, Morecroft 1988, Meadows 1980.). The objective of this book is to develop this broader role of System Dynamics diagrams as well as to explore the use of computer simulation models.

The use of System Dynamics diagrams to structure and analyse ill-defined situations can be considered as a free standing methodology, having much in common with the soft system problem solving methodologies recently developed as an alternative to science-based approaches (Checkland 1983, 1987, Ackoff 1978, Eden *et al*. 1979, Bryant 1989, Rosenhead 1989, Keys 1988). Alternatively, this extended use of diagrams can be considered as a front end addition to the conventional System Dynamics methodology.

The holistic and synergistic approach of System Dynamics also has much in common with other cross discipline methods of management, such as Business Policy (Gluek 1976) and Total Quality Management (Oakland, 1989).

Two characteristics of System Dynamics arising from its holistic view are worth highlighting at this stage. The first is its ability to generate structures which can be transferred to create insights in other systems. Spotting isomorphisms where others do not is often considered as the key to real intelligence (Sculley 1987) and System Dynamics provides a way of developing this skill. The second is its ability to help in identifying the counter-intuitive behaviour of systems. Often the implementation of new policies in systems results in unintentional side effects and System Dynamics provides a means of helping to determine what these might be.

DEFINITION AND OUTLINE OF THE SYSTEM DYNAMICS METHOD

There have been numerous definitions of System Dynamics. The one presented here is personal to the author and is intended to both combine previous definitions and to capture the broadening of the method which has taken place. This can be stated as follows:

> "A rigorous method for qualitative description, exploration and analysis of complex systems in terms of their processes, information, organisational boundaries and strategies; which facilitates quantitative simulation modelling and analysis for the design of system structure and control".

The methodology is very much embedded in the cybernetic or control paradigm and has been defined elsewhere (Coyle 1977) as 'that branch of control theory which deals with socio-economic systems'.

The author's definition is expanded in Figure 1.1, which provides a summary of the steps involved in the method and their purpose. The subject will be seen to comprise two separate phases which can be implemented in response to the identification of a problem or cause for concern. These are respectively, Qualitative and Quantitative System Dynamics and will be described in turn.

Qualitative System Dynamics	Quantitative System Dynamics	
(Diagram construction and analysis phase)	(Simulation phase)	
	stage 1	*stage 2*
To create and examine feedback loop structure of systems using resource flows, represented by level and rate variables and information flows, represented by auxiliary variables.	To examine the quantitative behaviour of all system variables over time.	To design alternative system structures and control strategies based on (i) intuitive ideas. (ii) control theory analogies. (iii) control theory algorithms. in terms of non-optimising robust policy design.
To provide a qualitative assessment of the relationship between system processes (including delays), information, organisational boundaries and strategy.	To examine the validity and sensitivity of system behaviour to changes in (i) information structure (ii) strategies (iii) delays/uncertainties.	
To estimate system behaviour and to postulate strategy design changes to improve behaviour.		To optimise the behaviour of specific system variables.

Figure 1.1 System Dynamics—a subject summary

It should be noted that, although in Figure 1.1 the steps of the approach are given as sequential, the method in practice, both within and between phases and stages, is an iterative procedure.

QUALITATIVE SYSTEM DYNAMICS

This phase of the method is based on creating cause and effect diagrams or system maps (known as causal loop or influence diagrams) according to precise and rigorous rules and using these to explore and analyse the system. The diagrams are developed with system actors to allow their mental models concerning system structure and strategies (and those of the environment of the system) to be made explicit. The word structure refers to the process and information structure of the system and is referred to as the information feedback structure of the system. Hence, System Dynamics models are often described as taking a feedback perspective of a situation.

It is an underlying premise of the subject of System Dynamics that the feedback structure of a system is a direct determinant of its behaviour over time.

The diagrams create a forum for translating barely perceived thoughts and assumptions about the system by individual actors into usable ideas which can be communicated to others. The intention is to broaden the understanding of each person and, by sharing their perceptions to make them aware of the system as a whole and their role within it; that is, to provide an holistic appreciation.

With experience it becomes possible to spot general feedback structures in systems and to develop insights from these perceptions. However, in many systems and, particularly, for the inexperienced analyst or system owner, it is necessary to develop feedback loop structures by a modular, stepwise approach. Considerable thought has been given to improving the ease of conceptualisation of System Dynamics Models in recent years (Morecroft 1982) and one such approach (based on Wolstenholme and Coyle 1983) will be described in this book. Apart from the guidance provided, this procedure also makes the translation of a qualitative model to a quantitative one much easier by identifying variables in a form compatible with simulation.

Once created, the diagrams can be used to qualitatively explore alternative structure and strategies, both within the system and its environment, which might benefit the system. Although comprehensive simulation is not advocated by the method at this stage, it is possible from a study of the feedback loop structure of the diagrams, to estimate their likely general direction of behaviour (say growth or decline). Further by using some of the experiences from the results of quantitative simulation modelling in other systems it is possible to apply guidelines for the redesign of system structures and strategies to improve system behaviour.

QUANTITATIVE SYSTEM DYNAMICS

The second phase of the subject is that of quantitative computer simulation modelling using purpose built software. This is the more conventional and traditional phase of System Dynamics and involves deriving with system actors the shape of relationships between all variables within the diagrams, the calibration of parameters and the construction of simulation equations and experiments. Although numbers are attached to variables during this phase, it should be stressed that the method is not aimed at accurate prediction or solutions. It is more concerned with the shape of change over time. Accurate prediction on the basis of past performance, assumes that the structure and strategies of the future will not be too dissimilar from the past. If the purpose of the model is to redesign structure and strategies, prediction must, by definition, be less accurate. Emphasis is on the *process* of modelling as a means of improving understanding. The idea being that such understanding will change perceptions and add to the ability of system actors to react better to future problems, that is, to make them more self-sufficient as problem solvers.

The power of quantitative System Dynamics has been significantly enhanced in recent years by the development of the desk-top computer and associated software. The creation of computer simulations of dynamic models has always been a significant factor in improving systemic understanding. This is because there is a severe limit in the cognitive ability of the human brain to process multi-variate problems without such help. Never before has computer power been so readily accessible and the potential this creates for experiential learning through questioning is enormous (De Geus 1988).

In systems which can be easily quantified it is possible to design specific control strategies and to provide accurate predictions of behaviour. In this sense the approach provides an advanced method to assist with the design and assessment of the information structure of systems. Recent advances in microcomputers have created a situation where information *retrieval* technology is at an advanced stage, however, the development of methods of assessing information *usage* via strategy has not been forthcoming. System Dynamics models can fulfil this latter role by allowing the testing of alternative system controls, based on alternative information sources. This role can provide an understanding of how information can be used to improve control and assist system actors in identifying information needs.

Control design methods used within System Dynamics have been developed vigorously in recent years (Mohapatra 1980, Sharma 1985 Keloharju 1983, Ozveren and Sterman 1989, Mosekilde, Aracil and Allan 1988). These approaches can be considered as providing a rear end addition to conventional System Dynamics and one of them, optimisation, will be considered in detail later in this book.

THE VALUE ADDED AT EACH PHASE OF THE METHOD

The value which can be added to understanding and insight into complex systems obviously increases with the depth of analysis which is applied, but so does the cost and effort required.

In general, qualitative analysis can often provide a significant level of understanding for a minimum investment of time and effort and is clearly appropriate when these resources are limited. The extension of the analysis to quantitative simulation obviously increases the level of understanding, but requires a much greater input of time and effort. In particular there is an overhead cost associated with learning to use the necessary computer software and hardware.

Further the value added at each stage of analysis per unit input, is a function of the type of system studied and the ability of the modeller. Qualitative analysis is appropriate for softer systems, which are difficult to structure and quantify and for those people with less numeric backgrounds.

It is precisely for the benefit of those people with less-quantitative backgrounds that the phase of Qualitative System Dynamics is developed in this book. There are a vast number of problematic situations where the major actors fall into this category and it is important that the powerful potential of feedback analysis is not denied to such people who, perhaps, need it most.

Quantitative analysis is appropriate for harder systems which are easier to structure and for modellers with a more numerical outlook.

It should be noted that the general shape of the relationship between value added and depth of analysis is also dynamic; particularly as improved computer hardware and software are introduced to improve the interface between people and computers.

AN OVERVIEW OF THE BOOK

The book will describe the process of applying the methodology of System Dynamics in the sequence outlined in this chapter. Chapters 2 and 3 concern Qualitative System Dynamics, using examples from a variety of system types. Chapter 2 presents the generic building blocks of the method, whilst Chapter 3 deals with describing the methodology and giving examples of its application.

Chapters 4 and 5 are concerned with the elements of Quantitative System Dynamics using computer simulation analysis. The material in these chapters is presented at a basic level to encourage those readers who have assimilated the ideas of Qualitative System Dynamics to sample the quantitative side of the subject. Concepts are introduced progressively to facilitate understanding of later case studies.

Three extensive case studies in Quantitative System Dynamics are then presented (Chapters 6, 7 and 8), using both the DYSMAP2 (Dangerfield and Vapenikova 1987, Vapenikova 1986) and STELLA (Richmond et al. 1988) software. The models underpinning these cases are presented in the Appendices. The detailed equation formulation of selected sectors of each model is described in the text.

Chapters 9 and 10 extend the depth of quantitative analysis to include Optimisation.

REFERENCES

Ackoff R.L. (1978) *The Art of Problem Solving*, Wiley, New York.
Bryant J. (1989) *Problem Management, A Guide for Producers and Players*, Wiley, Chichester.
Checkland P.B. (1983) O.R. and the Systems Movement: Mappings and Conflicts, *Journal of the Operational Research Society*, **34**, 661–675.
Checkland P.B. (1987) The Application of Systems Thinking in Real-World Problem-Situations: The Emergence of Soft-Systems Methodology, in M.C. Jackson and P. Keys (eds) *New Directions in Management Science*, Gower, Aldershot, 87–96.
Coyle R.G. (1977) *Management System Dynamics*, Wiley, Chichester.
Dangerfield B. and O. Vapenikova (1987) *DYSMAP2 User Manual*, University of Salford.
De Geus A.P. (1988) Planning as Learning, *Harvard Business Review*, March–April 1988.
Eden C., S. Jones and D. Sims (1979) *Thinking in Organisations*, MacMillan, London.
Forrester, J.W. (1961) *Industrial Dynamics*, MIT Press, Cambridge, MA.
Forrester J.W. (1968) *Principles of Systems*, MIT Press, Cambridge, MA.
Forrester J.W. (1969) *Urban Dynamics*, MIT Press, Cambridge, MA.
Forrester J.W. (1971) *World Dynamics*, MIT Press, Cambridge, MA.
Forrester J.W., A. Graham, P. Senge and J. Sterman (1983), *An Integrated Approach to the Economic Long Wave*, working paper D-3447-1, System Dynamics Group MIT Cambridge, MA.
Gluek W.F. (1976), *Business Policy*, McGraw Hill, New York.
Keloharju R. (1983) *Relativity Dynamics*, Helsinki School of Economics, Helsinki, Finland.
Keys P. (1988) System Dynamics: A Methodological Perspective, *Transactions of the Institute of Measurement and Control*, **10**, No. 4, July–September 1988.
Lyneis J.M. (1980) *Corporate Planning and Policy Design: A System Dynamics Approach*, MIT Press, Cambridge, MA.
Meadows D.M. *et al.* (1972) *The Limits to Growth*, Universe Books, New York.
Meadows D.M. (1980) The Unavoidable A Priori, in J. Randers (ed.) *Elements of the System Dynamics Method*, MIT Press, Cambridge, MA, pp.23–57.
Mohapatra P.K.J. (1980) *Part 1—Structural Equivalence Between Control System Theory and System Dynamics, Part II—Non-linearity in System Dynamics Models*, DYNAMICA, Vol. 6, pt. 1.
Morecroft J.D.W. (1982) *A Critical Review of Diagramming Tools for Conceptualising Feedback System Models*, DYNAMICA, Vol. 8, pt. I, pp.20–29.
Morecroft J.D.W. (1988) System Dynamics and Microworlds for Policy Makers, *European Journal of Operational Research*, **35**, 301–320.
Mosekilde E., J. Aracil and P.M. Allen (1988) Instabilities and Chaos in Non-Linear Dynamic Systems, *System Dynamics Review*, **4**, pp 56–81.
Oakland J.S. (1989) *Total Quality Management*, Heinemann Professional Publishing, London.
Ozveren C.M. and J.D. Sterman (1989). Control Theory Heuristics for Improving the Behaviour of Economic Models, *System Dynamics Review*, **5**, 130–148.
Richmond B., Peterson, S. and P. Vescuso (1988). *An Academic User's Guide to STELLA*, High Performance Systems Inc.
Richardson G.P. and A.L. Pugh (1981), *Introduction to System Dynamics Modelling with DYNAMO*, MIT Press, Cambridge, MA.
Roberts E.B. (1978) Strategies for Effective Implementation of Complex Corporate

Models, in E.B. Roberts (ed.) *Managerial Applications of System Dynamics*, MIT Press, Cambridge, MA, pp. 77–85.

Roberts N. *et al.* (1983) *Introduction to Computer Simulation: The System Dynamics Approach*, Addison-Wesley, Reading, MA.

Rosenhead, J. (ed.) (1989) *Rational Analysis for a Problematic World, Problem Structuring Methods for Complexity, Uncertainty and Conflict*, Wiley, Chichester.

Sharma S.K. (1985) *Policy Design in System Dynamics Models: Some Control Theory Applications*, Doctoral Thesis submitted to the Indian Institute of Technology, Kharagpur, India.

Sculley J. (1987) (with J.A. Byrne) *Odyssey: Pepsi to Apple*, Fontana, London.

Vapenikova, O. (1986) *The Development of DYSMAP2*, Proceedings of the 1986 International Conference of the System Dynamics Society, Sevilla, Spain.

Wolstenholme, E.F. (1982) System Dynamics in Perspective, *Journal of the Operational Research Society*, **33**, 547–556.

Wolstenholme E.F. and R.G. Coyle (1983) The Development of System Dynamics as a Methodology for System Description and Qualitative Analysis, *Journal of the Operational Research Society*, **34**, 569–581.

_____ Chapter 2

Qualitative System Dynamics I—The Components of Diagrams

INTRODUCTION

System Dynamics is centred on the use of diagrams as a medium for transmitting mental models and discussing change. This type of medium is useful because it provides a less ambiguous and more condensed form of communication than a written description.

There are many types of diagrams which have been used to describe systems, but there are few which are able to fulfil the role well. A prime prerequisite is to use only a small, sound and rigorous set of symbols or generic building blocks.

Such a set is available in System Dynamics and the purpose of this chapter is to introduce them. The elements of the subject are presented here in isolation since it is important that they are fully understood before describing how they are linked in methodological terms for model construction and analysis in Chapter 3.

The System Dynamics method is based on the premise that systems are composed of two basic components; process structure and information structure. There are also two generic building blocks which can be used to represent the structures. These are resource flows and information flows, respectively, which will be presented in turn.

CREATING PROCESS STRUCTURE

The processes within a system are not easily visible and in order to create a sound process perspective of a system it is usually necessary to stand back from the system, and view it over a sufficient time frame and at an appropriate level of aggregation. For example, if the concern is provision of homes for elderly people, it is important not to focus too finely on the needs of the individuals nor on the national total of elderly, but to view the total number of people in homes in one accounting unit, say a city, and the weekly or monthly flow into and out of these.

The System Dynamics approach to creating the process structure of systems is to recognise that the fundamental process in any natural or managed system is that of converting resources between states. The word resource here should be treated in its widest sense and could include material, people, cash, orders, goods, knowledge, etc. A state of a resource can then be defined as any accumulation of the resource which is relevant to the concern and, hence, purpose of the model. The states are alternatively known as system *levels* or *stocks*. They are the measurable quantities of any resource in a system at any point in time, and their dimensions are usually in resource units. If a photograph or static view is taken of the system, then its resource states will still be present and this technique provides a good way of identifying such variables. Figure 2.1 gives some examples of resources and their states.

Resource	State 1	State 2	State 3	State 4 etc.
Land	Wild	Cultivated		
Minerals	Undiscovered	Discovered	Exploited	Refined Stock
Productive Plant	Under Construction	New, Efficient Plant	Old, Obsolete Plant	Scrap Metal
Hospital Patients	Waiting for Treatment	In Hospital	Recuperating at Home	Back at Work
Scientists	University Based	Working on Research	Senior Staff	Retired
Labour	Untrained	Trainees	Trained	
Orders	Backlog	In Production	Satisfied	
Money	Bills Collectable	Cash in Hand		

Figure 2.1 Examples of resources and their states

The rate at which resources are converted between states is represented in System Dynamics by *rate* variables. Rate variables are control variables which directly increase or deplete resource levels and their dimensions are usually in units per period of time. That is, they control flows into and out of stocks. Rates can be considered as taking place instantaneously and, therefore, are not directly measurable.

The process structure of systems, as represented by resource flows made up of levels and rates, can be described by two types of diagrams. These will be referred to here as pipe diagrams and influence diagrams (alternatively known as causal loop diagrams). The alternative name for a 'pipe' diagram is a 'flow' diagram, but this name is avoided here because it conflicts with the word as used in 'resource flows'.

Figures 2.2a and 2.2b show pipe and influence diagrams for the same resource flow, representing patients flowing into hospitals and out into the community. Here, the resource is the people and the states chosen are patients in hospital and ex-patients in the community.

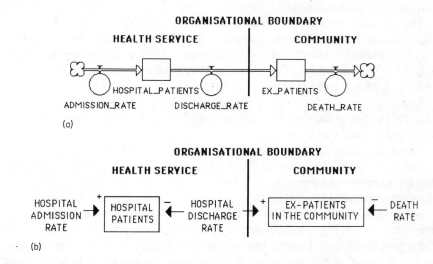

Figure 2.2 (a) Pipe diagram and (b) influence diagram for a resource flow

In the pipe diagram representation (Figure 2.2a) the resource flow is depicted by the double line (or pipe) connecting a source to a sink, both of these elements being represented by a 'cloud' symbol indicating an infinite availability of the resource outside the boundary of the model. The rates can be thought of as control valves which allow the resource to flow from the source into the level (e.g. hospital admission rate) and out from the level into the sink (e.g. death rate).

In the influence diagram representation (Figure 2.2b) the 'influences' of the rates on the level are specified using arrows and the sources and the sinks are omitted. Further, the polarity of the influences are specified and, in the convention used in this book, the levels are boxed.

To use influence diagrams comprehensively it is vital to have a thorough understanding of the concept of polarity. In general, if a change in the magnitude of the tail variable of an influence arrow causes a change in the magnitude of the head variable in the same direction, then the link is a positive one. Conversely, if a change in the magnitude of the tail variable of an influence arrow causes a change in the magnitude of the head variable in the opposite direction, then the link is a negative one.

For example in Figure 2.2b, an *increase* in the hospital *admission* rate would result in *more* patients in hospital and a *decrease* would result in *fewer*, hence, the direction of the influence is designated as positive. Conversely, an *increase* in the hospital *discharge* rate would result in *fewer* patients in hospital and a *decrease* would result in *more*. Therefore the direction of the influence is designated as negative.

The polarity attached to links in influence diagrams is important and facilitates analysis of composite models as will be seen in later examples.

It is important to realise that the direction of the resource flow in Figure 2.2b is (as in Figure 2.2a) from left to right even though some of the influence links point from right to left.

Resource flows can be physical (material) flows (for example the flow of money, people, goods, etc.) or non-physical flows (for example the flow of knowledge, motivation, etc.).

Physical flows can be represented as conserved or non-conserved. Conserved means that none of the resource can be lost or gained within the model. In terms of both influence and pipe diagrams, conserved flows can flow only between levels containing specified quantities of each resource.

More often than not, physical flows are represented as non-conserved. For example in Figures 2.2a and 2.2b the hospital admission rate could deplete a level representing the whole population. Hospital patients are, however, small in number relative to this and it is appropriate to consider the hospital admission rate as taking from an infinite source which is not explicitly modelled.

Non-physical flows are never conserved because resources like knowledge are never limited in quantity.

For the mathematically inclined, it is of interest to note that the process representation used is one of differential calculus, where a level represents the net integral or accumulation of the rate variables into and out of the level over a given period of time, that is, the area under the rate curve over that period.

For those familiar with entity life cycle diagrams as used as a basis for discrete simulation methods in queuing and congestion analysis, the rate variables used in System Dynamics diagrams can be considered as analogous to activities and the levels as analogous to queues.

DELAYS IN PROCESSES

One of the major factors contributing to system behaviour over time is that of delays. It is frequently the case that there is a lag between the start and finish of a resource conversion rate and this can be represented in an influence diagram by a letter 'D' linking the start and finish rates as shown in Figure 2.3.

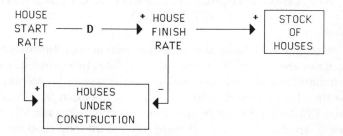

Figure 2.3   Influence diagram of a delayed resource flow

In general, in resource flows, levels can only depend on rates. Levels should never depend on other levels and, with the exception of delays, rates should never depend on other rates. The reason for the exception is that a delay is actually a hidden level, where the resource is held up and the quantity of resource which is delayed can be measured as shown in Figure 2.3.

MARKING ORGANISATIONAL BOUNDARIES

Once the process structure of a system has been drawn it is useful and important to superimpose any relevant organisational boundaries which exist within and between each resource flow. For example, in Figures 2.2a and 2.2b the resource of people flows through two different organisations: the health service and the community. The main purpose of marking such boundaries on the diagrams is to try to clarify which organisations or people control each rate variable in the process. In the example it is clear that the health service controls the hospital admission and discharge rates and that the community has no control over the numbers of ex-patients discharged into it.

Further, it may be necessary to draw any relevant departmental boundaries which exist within each organisation. If, for example, different departments or people within the health service control the admission and discharge rates, it is important to recognise this.

One of the most common reasons for problems encountered in lengthy processes in large systems is the number of different organisations controlling different parts of the process. If organisations do not integrate their control strategies with adjacent organisations, the processes will not flow smoothly.

CREATING INFORMATION STRUCTURE AND STRATEGY

It will be appreciated from the previous section that the magnitude of resource flows are controlled by the rate variables. In Figures 2.2a and 2.2b these variables have not yet been defined, in the sense that no causality has been specified for them and no arrows lead into them. In fact, the diagrams of Figures 2.2a and 2.2b can be described as open loop models. Creating information structure will convert these into closed loop models.

Two pieces of knowledge are required before the rates can be specified. The first piece of knowledge is which system states will be defined to have a causal effect on the rate. It should be noted that rates can only depend on levels, since these are the only measurable variables of a system. The second is what rule will be defined to specify the type of effect.

In managed systems the first piece of knowledge is the information chosen by the system owners and the second is the strategy by which to use the information. For this reason rate variables in managed systems are usually referred to as policy, strategy or decision variables.

Often the setting of rate variables involves defining targets (desired or objectives states) for levels and implementing strategies to eliminate any discrepancies between the target and actual values of the levels.

Composite, pipe and influence diagrams containing both resource and information flows and target states are shown in Figures 2.4a and 2.4b, respectively. The strategy represented in both diagrams is to fill the hospital capacity, that is, at any time the hospital admission or discharge rate will be changed to remove any discrepancy which exists between the capacity (target number of beds) available for patients and the actual number of beds filled.

The important idea in closed loop models is that information flows link knowledge about levels to rates and specify how the rates are to change in the future to change the quantities of the resources in the levels. In influence diagrams the direction of the change in magnitude of the information link is specified by the polarity of the arrows. In Figure 2.4b it is implied that as the number of patients in hospital increases (towards the hospital capacity), the admission rate must be reduced and/or the discharge rate increased.

In natural systems rates change over time even though they are not

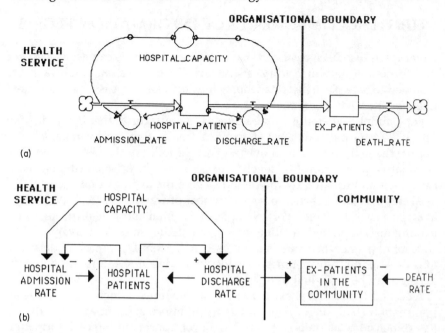

(a)

(b)

Figure 2.4 (a) Pipe diagram and (b) influence diagram for a combined resource and information flow

directly influenced by managerial action. They are affected by natural laws which often change in proportion to the magnitude of system levels, hence although they are not strictly information flows they can be depicted in a similar way. Such flows are referred to as behavioural flows.

Figure 2.5 shows an influence diagram representation of a combined resource–behavioural flow, where staff leave a company at a natural wastage rate outside the influence of management, which increases as the magnitude of the staff level increases and decreases with the average length of employment. The leaving rate could hence be modelled as the number of staff divided by the average length of employment.

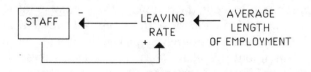

Figure 2.5 Influence diagram for a combined resource/behavioural flow

FURTHER COMPONENTS OF INFORMATION FLOWS

So far it has also been assumed that only resource levels can act as a source of information on which to base strategies. In practice, although rates are assumed to take place instantaneously they too can be measured if averaged over some period of time.

For example, when it is quoted by a car manufacturer that its production rate is 20 cars per day what it really means is that, on average, 20 cars per day have been produced over a given period, say the last 30 days. The cumulative production (the level) over 30 days which is 600 cars has been calculated and divided by the period to give the average rate.

Averaging can be carried out in a number of ways other than the straight average just calculated. For example, it is often more realistic to use a continuously updated moving or exponentially smoothed average. The concept of a smoothed rate is often used in System Dynamics models. The idea of smoothing over time is analogous to delaying information and an average can be modelled as a level.

A second type of level exists in System Dynamics, which is known as an information level, since it can only occur in information flows. This always has dimensions of units/time and is never conserved, since information may be created or destroyed. Figure 2.6 shows such a variable (average order rate) used in an order backlog model to regulate the production rate.

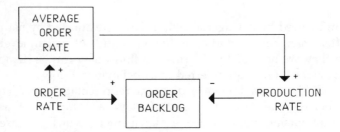

Figure 2.6 An example of a smoothed information level

It has also been assumed so far that levels feed information direct to rates, which subsequently modify resource flows to change levels. In practice information or behavioural flows can form extensive networks and it is usually appropriate to split them down into a number of steps. This is achieved by the introduction of a third type of variable known as an *auxiliary*. These variables are purely steps leading from levels to rates, with the aim of fleshing out the reality of the information flows.

For example, in Figure 2.7 orders on a particular company are cumulated into a backlog, which is decreased as the items ordered are produced.

However, customers do not perceive the company's order backlog directly, rather they respond to the 'lead time' quoted by the company and may reduce orders if this is excessive. This 'lead time' is effectively the order backlog divided by the average production rate, that is, the time to fulfil all outstanding orders at the current average production rate.

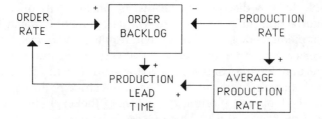

Figure 2.7   Example of detail in an information flow

The variable 'lead time' is technically unnecessary in modelling terms, since the effect of backlog and production rate on orders could be modelled without it. However, it is a real variable used by both the company and its customers and should be included in a model as an auxiliary variable. Further, it may be used directly in other parts of the company which are to be modelled.

FEEDBACK LOOPS

The major use of a System Dynamics diagram is to identify information feedback loops which have been created by linking resource and information flows. It is the analysis of such loops which facilitates understanding of how the processes, organisational boundaries, delays, information and strategies of systems interact to create system behaviour. The contribution of feedback loops to system behaviour depends primarily on whether they are positive or negative (that is, their polarity).

Influence diagrams facilitate identification of whether a feedback loop is positive or negative. The rule for this is that if the net effect of all individual influence links in a feedback loop is negative, then the whole loop is negative. Conversely, if the net effect is positive than the loop is positive. The net effect can be obtained by multiplying together the signs of the individual influence links (remembering that two negatives create a positive).

It should also be noted that a feedback loop must contain at least one rate and one level. Without integrating at least one rate into one level a loop cannot move forward through time and produce behaviour.

Consider first a negative loop. These are always target seeking (or control) loops and a number of examples of simple negative loops are given in Figure 2.8.

Figure 2.8a is a generic control loop, where the policy defining the rate variable might be to correct the gap or discrepancy between the actual size of the level which it feeds, and a target state for this. To create a rate variable from a discrepancy means, in practice, dividing by a time factor (or correction time), whose size will determine the rate at which the control is implemented. A small factor will result in a very fast, but possibly unstable correction and large factor in a very slow, but possibly more stable correction. Such a control policy is known in control engineering as proportional control. By tracing the effect of the strategy through a number of cycles of the loop and plotting the values of the level (and the rate) the behaviour of the loop unfolds. This process is known as simulation.

Figure 2.8b shows a specific classical negative feedback loop associated with a central heating system. If the room temperature (level of heat in the room) is below the desired temperature (thermostat setting), the central heating boiler will switch on to supply heat to the room to correct the discrepancy over a period of time. As the temperature of the room rises (the discrepancy between the desired and the actual states falls) the rate

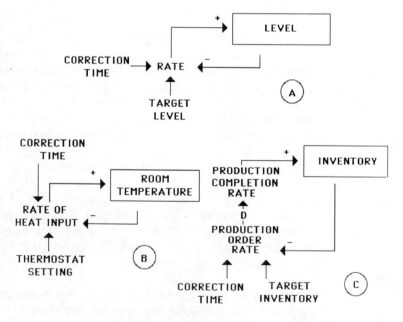

Figure 2.8 Examples of negative feedback loops

of heat input can be progressively reduced and the behaviour of the room temperature over time will be one of smooth transition to the desired level. This does, of course, assume the absence of exogenous (external) factors such as no heat loss from the room, no opening and closing of doors and no delay in raising the room temperature. The behaviour of the loop over time, as created using the System Dynamics computer simulation program STELLA, is shown in Figure 2.9 by the curve marked undelayed loop. This is a typical mode of behaviour from an undelayed negative loop.

Figure 2.8c shows a second specific negative feedback situation associated with a production–inventory system. This system contains a delay between production ordering and production completion. Any shortfall between target inventory and inventory at any time will result in increased orders. However, even though ordering will stop when inventory reaches its target, the inventory will continue to rise (and overshoot its target) due to work already in the production pipeline. Orders will be subsequently cut back and undershoot will occur. Figure 2.9 (marked delayed loop) shows the typical behaviour of such a delayed negative feedback loop which will oscillate around the target before reaching it. Other factors will cause negative loops to reach their targets in different ways. The important point to note, however, is that negative loops are target seeking.

Other examples of negative loops have been seen in Figures 2.4, 2.5 and 2.7. As already described, the two feedback loops shown in Figure 2.4b represent an attempt to control the hospital admission and discharge rates to change the numbers of patients in hospital over time towards the capacity of the hospital. For example, if the number of patients is below the capacity the admission rate will be increased. This increases the number of patients

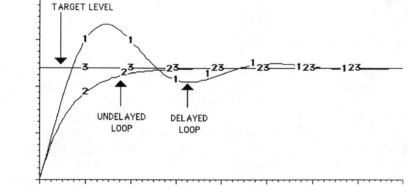

Figure 2.9   Classical modes of behaviour of a negative feedback loop

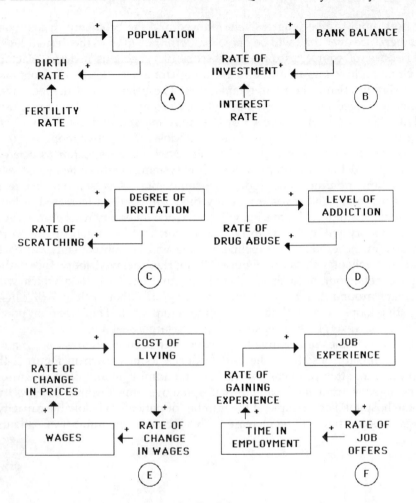

Figure 2.10 Examples of positive feedback loops

in hospital and reduces the hospital admission rate in future time periods.

In Figure 2.5 the staff leaving rate goes up as the staff level increases and this in turn decreases the level of staff. Since this is a behaviour response, there is no 'managerial' target involved for the staff level. In Figure 2.7 the order rate falls as the order backlog and production lead time increases and this reduces the order backlog. (Remember that the positive sign on the link between order rate and order backlog means that these two variables move in the *same* direction, that is, if the order rate is *reduced*, the order backlog is *reduced*.)

Positive feedback loops result in growth or decline rather than target seeking behaviour.

Examples of positive feedback loops are given in Figure 2.10. It will be seen that the processes represented are self-reinforcing. In Figure 2.10a increases in birth rate lead to increases in population, which in turn lead to increases in birth rate. In Figure 2.10b an increase in the rate of investment leads to a greater bank balance, which in turn leads to an increase in investment. The fertility rate and interest rate in these two figures act as 'gain' factors and the greater these are the greater the growth of the loops will be.

Undesirable positive feedback loops, such as those represented by Figures 2.10c, 2.10d, 2.10e and 2.10f, are often more popularly known as vicious circles or spiral effects. A Catch 22 situation is the name often given to a situation where it is impossible to break out of (Figure 2.10e) or into (Figure 2.10f) a vicious circle. These figures are also examples of feedback loops which are created between two different resource flows. Figure 2.11 presents the classical behaviour mode of a positive feedback loop over time.

Positive feedback loops can just as easily exhibit decline rather than growth over time and such loops are referred to as positive degenerate loops.

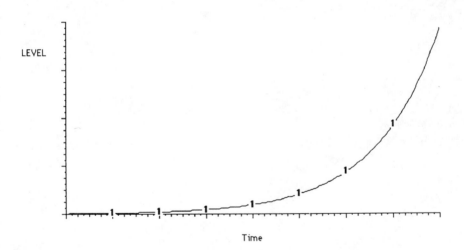

Figure 2.11 Classical mode of behaviour of a positive feedback loop

SUMMARY

This chapter has presented a brief overview of the generic components of system dynamics and how these can be assembled to create simple information feedback loops. It has also given an indication of the types of behaviour which can be expected to unfold by simulating or tracing round such loops over time.

The use of these building blocks to develop models is a theme extending right through this book. The ideas presented here will be strongly and progressively reinforced in later chapters. Chapter 3 deals with the general steps involved in creating and analysing models for specific purposes. These steps are then applied qualitatively. Subsequent chapters will deal with the development of quantitative models.

Chapter 3

Qualitative System Dynamics II—The Methodology and its Application

INTRODUCTION

This chapter describes the steps involved in the methodology of Qualitative System Dynamics and a case study is presented to demonstrate its application. The chapter also describes, via examples, the use of Qualitative System Dynamics to explore and test the likely effects of potential strategies prior to implementation.

MODEL CONSTRUCTION

Any system dynamics study should clearly be based on a defined cause for concern. Ideally, the concern should be specified in terms of existing, undesirable system behaviour and such a mode of system behaviour is often labelled as a reference mode of behaviour for the system. The definition of the concern is of great importance, since it dictates the shape and boundaries of the model.

System Dynamics models can be constructed in two basic ways, which in practice tend to be used together. They can proceed by identifying feedback structures (or loops) responsible for the reference mode behaviour of the system (the feedback loop approach) or by identifying specific examples of process, information, delay, strategy or organisation associated with the cause for concern (the modular approach).

Which of the elements is used as the starting point in the modular approach, depends on the purpose of the investigation and the type of system. For example, in a study originating from a reorganisation of managerial responsibility, it is logical to start with the new organisational boundaries and to study how the underlying processes of the organisation will be affected. Conversely, when studying the effects of a new information system, it is logical to start with the information flows.

The feedback loop approach obviously requires a reference mode of behaviour as a starting point and the creation of appropriate feedback loops, almost out of thin air, is often very difficult, particularly for the novice. The modular approach is an extension of the techniques already used to describe the components of the System Dynamics method in Chapter 2. This provides a very structured approach which leads to the early identification of variables and the construction of feedback loops and, hence, facilitates the construction of quantitative models.

The Feedback Loop Approach to Model Construction

Consider first the identification of feedback structure. It has been shown in the previous chapter that certain types of simple feedback loop create certain types of system behaviour. It follows, therefore, that if the existing reference mode of behaviour of a system is known, then it should be possible to infer the types of loops of which it is composed.

For example, if a system exhibits slow growth it is possible that there is a dominant positive loop present, but that this is being inhibited or constrained by a negative one. Similarly, a system which is declining slowly might be construed as being dominated by a positive degenerate loop, which is being controlled by the imposition of a number of negative effects.

Alternatively, a stair step or signoidal type of growth pattern might be indicative of a positive loop which is dominant most of the time, but occasionally overwhelmed by a negative influence. Similarly, an s-shaped growth might indicate a shift in dominance from an initial positive growth loop to a negative control loop.

Feedback loops can be thought of as the forces within a system which combine to pull it in different directions of evolution.

This method of model construction evolves by identifying as many loops as possible and linking them together. Once each feedback loop has been identified it is useful to identify the variable types of which it is composed, that is, the rates, levels and auxiliaries. This can be a difficult task but assists with understanding how the loops contributes to behaviour, as well as being a necessary step in moving towards a quantitative model. Further, its components of process, information, strategy and organisational boundaries should be recognised.

The Modular Approach to Model Construction

The second method of model construction is almost the opposite to the one described above. In the absence of a reference mode, it is necessary to start with one or two key variables associated with the cause for concern and to try to relate examples of process, information, delay, strategy or organisation to these. The particular variant of the method to be described here begins by creating the process structure.

The procedure for creating the process structure of a system is, firstly, to identify system resources associated with the key variables and, secondly, to proceed as described in Chapter 2. That is, to identify some initial states of each resource and thirdly, to construct resource flows for each resource, containing relevant resource states and their associated rates of conversion.

A resource flow must contain at least one resource state and one rate. If, as in Figures 2.2a and 2.2b, more than one state of a resource is involved the resource flows can be cascaded together to produce a chain of resource conversion.

Delays in processes and organisational boundaries should be identified and superimposed on the diagram at this stage.

The information structure within a resource flow is then created by identifying which information is used in determining each rate variable.

In a practical model, the procedure of creating process and information structure would be carried out for a number of resources in turn. In addition to the information–behavioural links created between levels and rates *within* each resource flow, similar links should be identified which exist *between* different resource flows.

It is the superimposition of the information feedback structure on the process structure of a model which creates the feedback loops of the model.

Figure 3.1 shows a two-resource model created from Figure 2.4b by considering hospital capacity as a level variable rather than a constant target. This extended model highlights the fact that responsibilities for changing hospital capacity lie with a different set of people from those controlling hospital admissions or discharge rates.

In general, once started the modular approach to model construction can be developed to explore the system and its environment. It is often useful to begin the process of diagram conceptualisation using one or two major resources and a small number of levels at a high degree of aggregation and with a clear time horizon (that is, the time scale of change of the concern) in mind. Even at this stage it helps to think about the order of magnitude of the variables involved. For example, it would be foolish to consider levels in the same model which had absurdly different magnitudes or rates which ranged from microseconds to years.

As the model develops by the superimposition of information–behavioural links on process structure, new resources or states relevant to the concern can often be identified or existing resources or states eliminated. When adding new resources or states it is necessary to reiterate the whole

Table 3.1. Summary of the steps in the modular approach to System Dynamics model creation and development

1. Recognise the key variables associated with the perceived cause(s) for concern. Where possible, obtain data on the behaviour of these variables over time and define a reference mode for the existing system behaviour over a suitable time horizon.

2. Identify some of the initial system resources associated with the key variables.

3. Identify some of the initial states (levels) of each resource. These initial states should be defined at a reasonably high level of aggregation.

4. Construct resource flows for each resource, containing the identified states and their associated rates of conversion. Include any significant process delays in the resource flows. (A resource flow must contain at least one resource state and one rate.)

5. If more than one state of each resource is involved cascade the resource flows together to produce a chain of resource conversion or transfer, alternating the levels and rates.

6. Within each resource flow identify organisational boundaries, behavioural–information flows and strategies by which the levels influence the rates. Include any significant delays in the information flows.

7. Identify similar organisational boundaries, behavioural/information flows and strategies *between* different resource flows. For complex situations this should be carried out for small groups of resources at a time within a defined theme and the resultant diagrams reduced to produce the simplest representation possible, consistent with relating the key variables of the investigation.

8. Identify any new states of existing resources, or new resources, which affect the variables created and add these to those identified in steps 2 and 3.

9. Reiterate if necessary.

process to investigate how the new elements intermesh with the old. For example, an additional resource in Figure 3.1 might be that of finance which relates directly to the variables affecting capacity changes. That is, new money would be needed for new hospitals and money could be saved by closing hospitals when they cannot be properly maintained or staffed.

The modular approach to model creation is iterative and geared to focusing attention on the best compromise between the degree of resolution of the model and its size. This usually involves expanding the boundaries of the model initially and then progressively contracting them.

The ideal outcome of the conceptualisation process is a model which

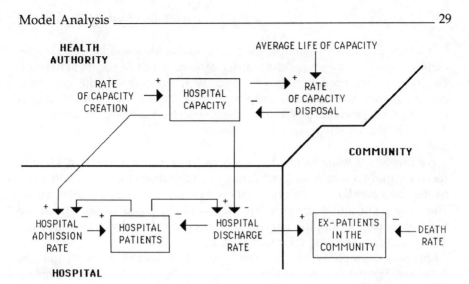

Figure 3.1 An example of a model with two resource flows

captures all the salient features of factors affecting the cause for concern in the simplest and most compact way.

The whole procedure can be described as exploring the system and its environment diagrammatically and a summary of the steps in the modular approach to model creation and development is presented in Table 3.1.

MODEL ANALYSIS

Once finalised System Dynamics diagrams can be used to provide a range of insights into system performance. One of these is to provide a clear description of the processes at work in the system. This should not be underestimated as perceptions of most human activity systems (that is, those constructed and operated by people) tend to centre on spheres of managerial responsibility and compartments of activity rather than on the processes for which the system exists.

The primary use of a System Dynamics model is, of course, to infer the behaviour of the system from the system structure and the reasons for this (that is, understanding 'what is'). This can then lead logically to system redesign (that is, designing 'what should be').

Behaviour of the model over time can be qualitatively assessed. This is achieved by identifying major information feedback loops and by tracing out the effects of changes to specific rate variables round the variables of the loops. The behaviour can then be compared with the reference mode of behaviour of the system to provide some degree of confidence in the validity of the model.

It is then possible, by manipulating information links and hence strategies within a system, to change its feedback structure and hence its behaviour. The type of manipulation to be undertaken depends on the objectives defined for behaviour. For example, if fast growth is the aim, then it is important to promote positive feedback effects. However, if arresting a decline is the target, then it is important to promote negative feedback loop control.

An important issue in moving towards improved control is to identify specific variables which are performing in an unsatisfactory way. If these lie within the boundary of an organisation then scope to control them exists. This may seem rather obvious but, all too often, organisations spend too much time responding to or trying to influence external variables and too little time on controlling internal variables

One way of implementing control is to define desired states for internal variables. Simple control strategies can then be introduced to eliminate discrepancies between the desired and actual states. In some circumstances, it may be necessary to introduce additional resource flows to implement control (controlling resources). For example, recruitment of additional categories of employees.

Table 3.2. Summary of the steps for model analysis using Qualitative System Dynamics

---

1. Isolate the major feedback loops in the model, whether arising intuitively or from the modular approach to model construction.

2. Assess the general mode of behaviour of the individual loops and the whole model over time arising from the strategies contained with them. This can be achieved in simple cases by determining the polarity of each feedback loop or in more complex cases, by tracing round each loop the effect of a change in one of its constituent rate variables. Check if this mode of behaviour is consistent with any reference mode available for the system.

3. Identify the rate variables within each loop which are available to be controlled, that is, those which are within the boundaries of the organisation trying to implement the system control.

4. Identify, possible ways to control these variables. For example, by defining target states for them or by linking them to information sources (levels) elsewhere in the model and specifying appropriate strategies by which to use the information.

5. Assess, as in 2, the general model of behaviour of the model arising from any new feedback loops created in step 4.

6. Reiterate from step 3, if necessary.

The underlying concept behind system dynamics analysis is that only by understanding with system actors how the elements of a system combine holistically to create its performance, can sensible and lasting change be made. It is often the case in practice that actions which are taken to improve system performance are piecemeal and restricted only to affecting the obvious symptoms of problems. Such change can at best be only of temporary advantage and at worse be detrimental.

A summary of the steps involved in applying Qualitative System Dynamics to model analysis are given in Table 3.2. An expanded list of these steps can be found elsewhere (Wolstenholme 1987). An example will now be presented to demonstrate the full use of the approach for developing a model, for using it to provide a comprehensive hypothesis of a perceived problem and for identifying appropriate solutions.

A CASE STUDY IN MODEL CONSTRUCTION AND ANALYSIS—THE ESS CASE

The Problem

This business problem concerns a small company (ESS) specialising in expert system software. Its initial product (Bradsoft) on which it was founded sold well on a worldwide basis with income from both sales and long term after-sales support contracts. These covered client training, problem solving and assistance with integrating the product into the client organisation. This success encouraged the company to invest profits into developing other similar products.

Sales and profits of Bradsoft have, however, recently fallen and certain key customers have ceased to use the product. Further, the turnover of staff at ESS has increased, resulting in training problems.

The management of ESS feel that most of the problems which have arisen are as a result of external factors. The loss of customers has been put down to the poor quality of client's staff and their inability to absorb new technology. The loss of sales is seen as due to competitors or as a function of the product being ahead of its time for a large section of the potential market. The loss of staff is seen as inevitable for a market leader.

The company intend to invest more heavily in expanding the product range by taking on additional product development staff. Their intention is to develop more frontier products to stay ahead of the field and to spread the risk.

The problems of ESS may well be due to outside factors and its lack of understanding of the market and it is important that better information of the market be assembled. In fact, such a focus would be the typical marketing or business policy approach to the situation described. What is emphasised in System Dynamics is that focus should equally be placed on the internal workings of the company. It is these workings which can be directly controlled and the initial hypothesis in this subject is always that a large proportion of problems are self inflicted. In other words it is important to explore the internal interactions between process, organisation and strategy which might lead the company to shooting itself in the foot.

Identifying Resources, States and Resource Flows

In order to explore the problem more thoroughly an influence diagram will be constructed. The key variable of concern is profit and the reference mode for this is of an initial rise followed by a fall. The initial resources and their states inferred from the problem description, together with their resource flows, are shown in Figure 3.2. The thinking behind the construction of Figure 3.2 will now be discussed.

Four resources of interest to the problem can be identified from the problem description. The first of these is products, which play a significant part in the company's intended development strategy. The second is money, which is both the key cause for concern and, potentially through investment, the road to salvation. The third is people, or more specifically staff, who are necessary for both product support and development. The fourth is customers.

Two relevant states for the resource 'products' are those 'under development' and 'at market' and a resource flow linking these is constructed in Figure 3.2. Products must be developed over a period of time before attaining the saleable state of being 'at market'. Those in this development pipeline can be considered as 'under development'.

An initial relevant state for the resource of money, is that of profit. This is increased by revenue and depleted by sales and investment, as shown in the resource flow in Figure 3.2.

Two states have already been identified for the resource of staff. These are product support and product development staff. Two structurally equivalent resource flows are shown for these states in Figure 3.2. Both involve recruitment rates, training rates, training delays and leaving rates.

Finally, for the resource customers, the company is chiefly concerned with those in the state of 'active customers'. A resource flow for this centres around the ideas of generating customers by sales and maintaining them as active by minimising the customer loss rate.

Having developed the ideas in Figure 3.2, the next step is to try to link the individual resource flows together into a composite structure, which will start to explain the behaviour of the company's profit.

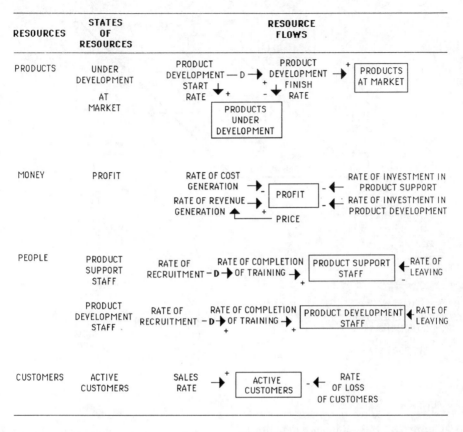

Figure 3.2 Resources, states and resource flows for the ESS case

A start will be made by constructing the processes and strategies within the company, which combined to promote the initial growth and then superimposing the company's current and proposed strategies.

An Initial Composite Model

Figure 3.3 shows a composite influence diagram, which links together the resource flows of Figure 3.2. The sales rate of the single product (Bradsoft) adds to the number of active customers using the product. The revenue generation rate, which cumulates into profit, is created both from sales and from active customers paying maintenance contracts. The initial profit growth of the company stemmed from these processes.

The early strategy was to use profits to recruit and train product support staff whose activities reinforced the growth rate by maintaining the base of active customers. In other words a positive feedback loop existed, which

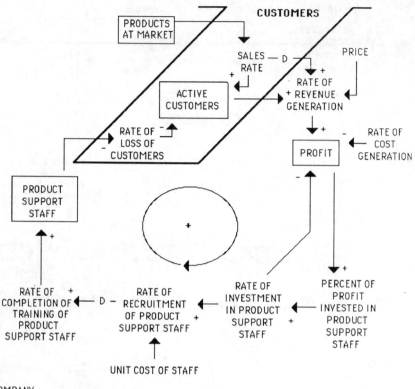

Figure 3.3 ESS—influence diagram 1

allowed the product support function to grow as the number of customers grew.

Linking together the resource flows of Figure 3.2 to create Figure 3.3 raises the possibility of other additional interactions and decisions must be made as to whether these should be incorporated into the model or assumptions made to exclude them. For example, product support staff will introduce additional costs to offset the revenue which they generate from reducing the loss of customers. Further, it might also be argued that both sales and the rate of loss of customers are a function of price. The management of ESS felt that price was not a significant factor and could be treated exogeneously to the qualitative model. Additionally, whilst accepting that product support staff directly affect costs, they were of the view that this was small relative to the contribution of these staff to revenue creation.

A Second Composite Model

Figure 3.4 shows an expanded influence diagram capturing the more recent strategy of ESS, which was to try to expand its number of products. The expansion was achieved by investing strongly in product development via the recruiting of product development staff. As shown in Figure 3.4, increasing the rate of investment of profits in product development does create a positive product development loop (A). However, such investment must decrease the rate of investment in supporting the current product. This creates a negative product support loop (B), highlighted in Figure 3.4 by the sequence of thick influence arrows. (Note that this loop contains three negative links and its net effect is, hence, negative.)

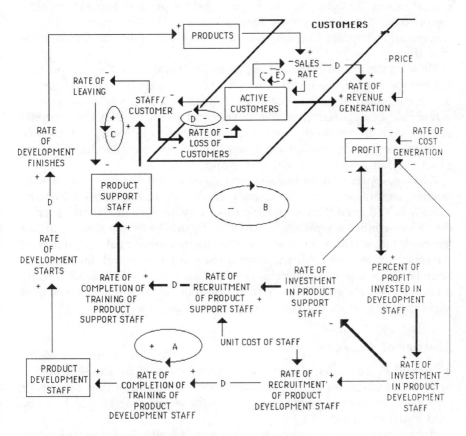

Figure 3.4 ESS—influence diagram 2

As more product development takes place the number of staff supporting each customer of Bradsoft is reduced. The ratio of product development staff to customers is defined as a new variable in Figure 3.4. It is a ratio of two levels and is added to the product support loop in Figure 3.4 as a useful representation of the quality of product support. As the customer load of each product support person increases, it is likely that there will be an increase in the leaving rate of these staff via the positive feedback loop (C) and further deterioration in product support will take place.

The reduction in product support is likely to be masked initially by individual staff increasing their effort and workload, but eventually might well contribute to customers being dissatisfied with the length of time it takes to install and operate the complex piece of software. Some may, in fact, abandon the product, which will ease the product support problem temporarily, via the negative feedback loop (D). As fewer customers successfully integrate the software sales may well fall, via the negative loop (E).

These effects are all contained in Figure 3.4. If and when the next product reaches the market, there will be a very welcome surge in sales and profits, but in the longer term the existing effects will be drastically compounded.

It should be apparent from the above that it would be disastrous for ESS to undertake further product development at this stage, since it is incapable of supporting its existing product, has less money to invest due to falling profit and has a lengthy product development delay.

What is apparent from the above analysis is that the company has no internal control. It is merely reacting to the revenue generated from its customers and even then employing this badly. In System Dynamics terms the lack of control is strongly indicated in Figure 3.4 by the absence of any feedback loops which lie wholly within the organisational boundaries of the company. All the feedback loops discussed pass through the customer boundary and depend on variables such as 'sales rate', which is not directly under the control of ESS. Such an observation highlights the purpose of superimposing organisational boundaries on diagrams.

Redesign of Strategies

By providing an understanding of the problem faced by ESS, Qualitative System Dynamics can provide a basis for designing appropriate changes or controls. A number of points emerge from the above analysis which can be used to define controls.

Firstly, the analysis indicates that customers are effectively the integration or accumulation of the sales rate and, hence, to maintain a given ratio of product support staff to customers the number of product support staff must be increased in proportion to the cumulative sales.

Secondly, it highlights ESS's unrealistic assumption that there is a need to support customers indefinitely. It would be useful to define an appropriate time span for product support.

There is an interesting analogy here between the number of product support staff in ESS and the number of spare parts required to be produced

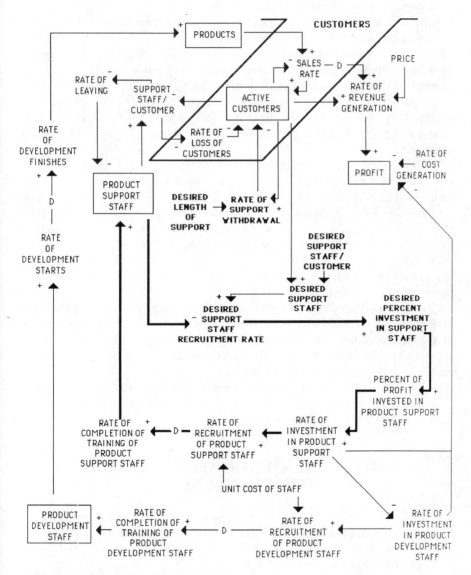

Figure 3.5 ESS—influence diagram 3

by many large manufacturing organisations. It is often not realised in the latter type of company that spare parts must be produced in proportion to sales of new machinery and, indeed, stocked for the estimated period of life of those machines.

Thirdly, a control mechanism should be devised to control the allocation of investment between product support and product development.

In general, as previously illustrated, the design of control for a system requires the identification of variables which are performing badly and the creation of measures to improve the control of these. In the case here, from a project support point of view, such a variable is the ratio of product support staff to customers, for no apparent control yet exists for this variable. Control could be introduced by defining a desired state for it at, say, a value which might prevent customer loss. In order to implement the control, new information structure must be superimposed on the model and, eventually, on the company. This is shown in bold type in Figure 3.5 and will now be described.

A knowledge of the actual number of customers and the desired support staff per customer, enables calculation of the desired number of support staff needed. A comparison of this with the actual number of support staff then enables (in conjunction with an appropriate strategy), the calculation of a desired rate of recruitment of support staff and to the calculation of the desired percentage of available investment required for support staff. This will, in turn, regulate the actual percentage of profit invested in development and product support, which is one of the existing model variables.

This control introduces a negative feedback loop *within* the boundary of ESS which is highlighted by the sequence of thick influence arrows in Figure 3.5. The control will ensure that an adequate level of product support is always maintained. The consequence should be a more stable customer base and an improvement in sales, revenue and future product development, whilst ensuring a lower turnover of product support staff.

Figure 3.5 also introduces the concept of product support withdrawal, based on a desired length of support.

THE TESTING OF STRATEGIES PRIOR TO IMPLEMENTATION

The previous example concerned the logical development of strategies for ESS, arising from an understanding of the company and its situation. Often, however, strategies are implemented in practice without careful design or without exploring their likely effects.

It has often proved the case that well intentioned strategies applied in one sector of a system can fail in practice, or even lead to unintentional side effects, due to reactions in other parts of a system which are provoked by them. Identification of such 'counter intuitive' outcomes of strategies form one of the major purposes in undertaking System Dynamics studies and Qualitative System Dynamics can be deployed to assist this process.

A number of examples will be presented which start from the point of analysis of a system at which it is considered that a viable control procedure has been defined and Qualitative System Dynamics will be used to try to understand how, in practice, such control might result in having the opposite effect on the variable of concern to that intended.

In System Dynamics terms this means translating the intended control into negative feedback terms, and then attempting to identify how the implementation of this negative loop may in fact result in the formation of a positive loop which becomes dominant.

The General Case

The general situation described is shown in the influence diagrams of Figures 3.6a and 3.6b. Figure 3.6a captures a state of a resource stream which is causing concern and attempts to control it by introducing a controlling resource. The *strategy* is to introduce the controlling resource at a given rate and the *intended effect* of the negative loop control is that the level of the controlling resource will reduce the rate of change of the state of the resource of concern. It is, however, suggested in Figure 3.6b that increasing the level of the controlling resource may provoke a *response* in some other part of the system or its environment, which via other resources may feedback to have an *actual effect* on the rate of change of the variable of concern which is positive. The purpose of analysis here is to try, in specific cases, to identify such unintentional response paths.

Children in Care

The first specific example of the phenomenon involves children in local authority care. Figure 3.7a shows a situation where there is concern about the rate of referral of children to care, apparently as a result of child abuse, and where the corrective *strategy* is considered to be to recruit more social workers.

The hypothesis underlying this *intended effect* in that additional social workers in the field would help to eleviate the necessity of bringing children into care, by resolving problems in individual households. Rather than being reduced, however, the rate of children entering care has, over recent years, continued to increase despite the employment of more social workers. Whilst it is easy to say, and possibly true, that the underlying reason for

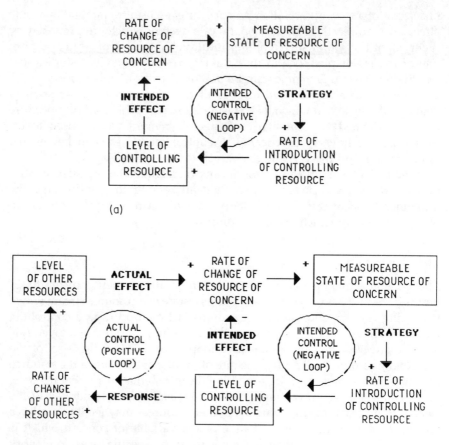

Figure 3.6 (a) General model of control and (b) general model of control and its counter intuitive effects

the continued increase is due to external factors, it is of interest to postulate how some of it may be linked to the implemented strategy. Figure 3.7b gives an explanation of how the level of social workers might itself create the paradoxical counter intuitive behaviour.

First of all it is necessary to observe that the actual rate of increase of children entering care is only a symptom of the real problem. This is stated in Figure 3.7a to be child abuse and it is important to define this variable more explicitly. A convenient way of proceeding is to introduce abused children as a resource stream as shown in Figure 3.7b. Here, the real rate of abuse creates a stock of cases of abused children and the rate of discovery of these transfers them to a stock of reported cases of child abuse. It is this

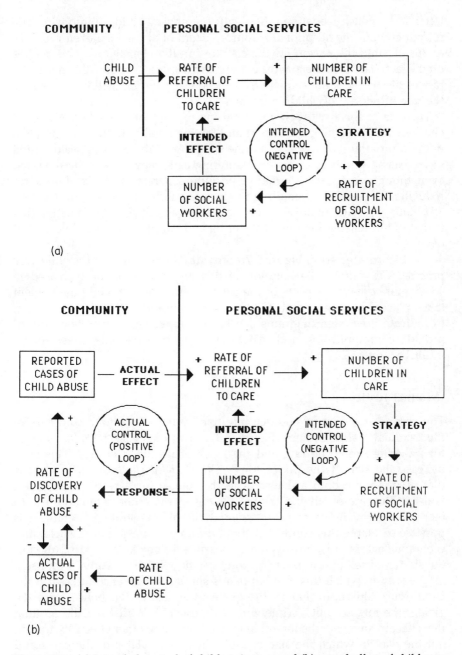

(a)

(b)

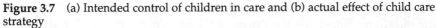

Figure 3.7 (a) Intended control of children in care and (b) actual effect of child care strategy

latter stock which feeds the rate of entry of abused children into care. The introduction of the additional resource stream of abused children provides an insight into the reason for the counter intuitive behaviour. This centres on the fact that more social workers, create a *response* which will increase the rate of discovery of abused children and the *actual effect* will be to increase the rate of referral of children to care.

The role of social workers in increasing the rate of discovery of abused children, may well be more important than their ability to treat the problem in the community. It suggests that the symptom of the problem, as reflected in increasing numbers of children entering care, may have to be tolerated even though costly or, alternatively, quite separate strategies found for reducing the number of children in care. The influence diagram of Figure 3.7b can assist with identifying those alternatives. An obvious consideration is that of influencing the rate of fostering and adoption to take children out of care.

The picture created in Figure 3.7b provides an overview of the child care process. The degree of resolution of this diagram could be increased to cover more detailed aspects of the process, such as how child assessment takes place. From the point of view of the objective here, it is sufficient to indicate how such diagrams can be developed in a controlled way to provide communication and insight into the possible side effects of an intuitive strategy.

Criminal Justice

The second example of counter intuitive system behaviour comes from the criminal justice system. Figure 3.8 shows one very popular *strategy* for reducing the crime rate and the stock of unsolved crimes. This is to increase the size of the police force. The *intended effect* of the control is that of providing both a deterrent to the rate of increase of crime and a boost to the rate of solving crimes. However, the crime rate continues to increase despite numerous implementations of this strategy. Again, it is possible to blame this continued trend on underlying political, social and economic factors. It is, however, also possible to look at the contribution of the strategy itself and to try to hypothesise how it may feed back on itself.

One way in which this will happen is similar to that encountered in the child abuse situation, that is, the greater the size of the police force the greater the rate at which crime will be discovered. Whilst not discounting this effect in any way, the intention here is to explore other effects by tracing out the way in which the size of the police force relates to the court and prison systems. Figure 3.8 describes the hypothesis and encompasses four organisations. These are the police, the community, the court system and the prison system.

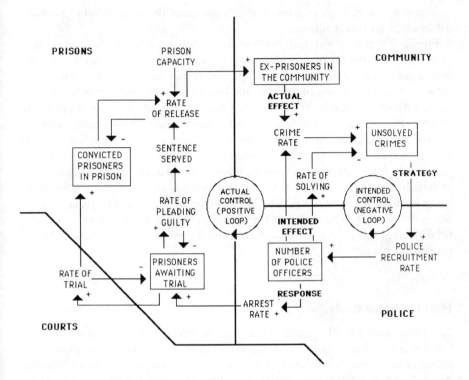

Figure 3.8 An example of control and counter intuitive effects in the criminal justice system

It is suggested in Figure 3.8 that a *response* to an increase in the size of the police force will be to create an increase in the arrest rate and, hence, in the number of people in prison awaiting trial. The rate at which cases can be tried depends on the capacity of the courts. If this is already stretched, one way of coping with the situation is to try to achieve a bypassing of the courts by encouraging accused people to plead guilty in return for a reduced sentence and early release. This sequence of influences provides one potentially positive *actual effect* on the crime rate variable, via the number of ex-prisoners in the community having served nominal sentences. The creation of such informal strategies and routes through systems always occur when systems approach their capacity and they can be thought of as safety valves. The danger is that such strategies often become part of the system and remain unquestioned unless a comprehensive investigation is carried out.

Even if the rate of trial and conviction can increase, however, a similar over capacity situation occurs in prisons. This again results in enormous pressure for early release; a route which provides the second major positive *actual effect* on the crime rate.

Figure 3.8 traces out the process by which people in the community pass through the police, courts and prison system, and back into the community. The total system through which they pass is an enormous one which transcends many organisational boundaries and sub-boundaries. The most important insight from the diagram is that an increase in capacity in just the police sector will do little more than create congestion in courts and prisons. To be effective, capacity increases must be applied simultaneously throughout the system. This is a good example of a very commonly found dynamic problem, where the variables controlling the rate at which resources enter each sector of a system, are outside the control of those sectors. The only way that each sector of the system can cope with increases in their input rates, is for them to increase their output rates. The total throughput of the whole system then increases with all the additional problems this entails.

Health Service Costs

The third example of counter intuitive behaviour comes from the Health Service and is captured in the influence diagram of Figure 3.9. In response to rising costs a Health Authority intends to implement a strategy of hospital ward closures, particularly geriatric wards. The *intended effect* is to have a controlling or negative feedback effect on costs. The resources of interest are those of capacity and costs and the intended control is captured by the negative feedback loop on the left of the diagram.

One of the effects of reducing hospital capacity is, obviously, to increase the rate of discharge of patients from hospital wards, which was modelled in Chapter 2. The process effect of the ward closure strategy is that people 'flow' into the community across an organisational boundary and this means that some other agency (the Local Government in this case) must accommodate the displaced people. Although some funding is provided from the Health Service to Local Government to assist with the transfer of people, the direct demand on homes for the elderly will increase as will the indirect demand for provision of assistance to those in the community, hence, Local Government costs, which are a fourth resource, will rise.

The *response* of Local Government to cost increases is likely to be an increase in Local Government charges on the community. Although this will be spread over numerous payees, one of the major ones will be the Health Authority, hence, as shown in Figure 3.9, the *actual effect* will be to create a positive feedback loop around the negative one, which will reduce the magnitude of the cost savings to the Health Authority.

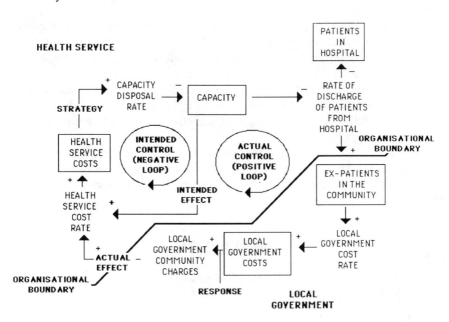

Figure 3.9 An example of control and counter intuitive effects in health care

The question in practice, in this example, is whether the magnitude of the cost savings in the Health Authority will outweigh the additional community charges levied. This is something that requires secondary analysis. The important contribution from Qualitative System Dynamics is the recognition that there will be an adaptive reaction from outside the Health Service to the ward closure strategy.

At the time of writing, consideration is being given to handing over full, direct, financial responsibility for the elderly to Local Government. This will give Local Government much greater control over the rate at which patients requiring community care are discharged from hospital. In System Dynamics terms, this will have the effect of moving the organisational boundary in Figure 3.9 from below to above the variable 'rate of discharge of patients from hospital'.

SUMMARY

This chapter has presented the whole process of Qualitative System Dynamics from both a general and specific point of view.

It must be emphasised that the approach is often more iterative than described here. The thinking generated in the early stages of model construction often requires many resources and states to be included and discarded and many potential links between resource flows to be considered. The purpose of the enquiry, the relevant level of aggregation of the resource states chosen and the time horizon of the model should always be kept in mind. Contact with the owners of the system being modelled is vital to create a full debate on the choices to be made around these issues. A good model evolves from working directly with people experienced in the system under study.

This dialogue is also vital at the system redesign state. Implementation of change actually begins by involving system owners in the design process and establishing analysis and understanding of alternative proposals for change.

REFERENCE

Wolstenholme, E.F. (1987) in J.S. Berry, (Ed.) *Mathematical Modelling, Methodology Models and Micros*, Ellis Horwood, Chichester.

Chapter 4

Quantitative System Dynamics I—An Overview of the Quantification, Simulation and Use of Models

INTRODUCTION

Chapter 3 demonstrated some of the ways in which Qualitative System Dynamics could be used to structure and analyse system problems. This chapter embarks on how such diagrammatic models might be converted into quantitative models to explore system behaviour and control more rigorously.

It should be stressed that the prime purpose of this chapter is to communicate the thinking behind quantification and to formalise the ideas of simulation. The basic structure of model equations, using the DYSMAP2[1] simulation language are described and other software introduced. An overview is given of the concept of model validation and the use of quantitative models for system redesign.

Detailed use of the principles described, particularly equation formulation will be introduced in later case studies. It is intended that this chapter should be read by those unfamiliar with quantitative simulation methods who would like to know about how such models can be developed.

QUANTIFICATION OF MODELS

The first step in creating a simulation model is to attempt to quantify the relationships suggested by the links of a diagrammatic model. Many of these will be automatic, in that level equations will be simply a function of their input and output rates. Problems can, however, arise in creating the equations for auxiliary and rate variables. Some of the relationships for these will have been in the mind of the experienced modeller at the model conceptualisation stage and diagrams possibly created with the ease of quantification in mind. In general, model relationships can be justified in a number of ways depending on the type of model. These include direct observation, accepted theory, hypothesis, assumption, belief or statistical evidence.

As modelling moves from the hard to the softer areas of the system spectrum so, obviously, quantification becomes more difficult. Ultimately, however, it must be remembered that most models are created for a purpose by an individual group to provide insight. Therefore, as long as there is agreement between participants as to the relationships, then the model satisfies its purpose. In fact, one of the fundamental purposes in modelling is to expose the assumptions within systems and a detailed analysis of relationships within models forces attention on this issue.

Two strategies can be employed to cope with any relationship which cannot be justified on firm grounds. Firstly, a tentative relationship can be constructed and kept fixed throughout all experimental work on the model. This approach can be justified in that, although it may not be possible to generate experimental results against an absolute standard, the relative results between model experiments will be consistent. Secondly, a range of tentative relationships can themselves be the subject of experimentation.

It is always important to try to quantify all aspects of a model, even if some of these have to be on a normative scale, since one of the major axioms of the approach is that the behaviour of whole systems is not predictable from the behaviour of its individual components.

Further, although relationships should be specified as accurately as possible, System Dynamics models are more concerned with capturing the structure and policies of systems and with the *mode* of behaviour of the whole system, rather than accurate prediction. It is considered that the *shape* of relationships is more important than their absolute, statistical accuracy and that in any holistic approach accuracy must often be sacrificed in order to remain problem orientated.

THE SIMULATION CONCEPT

It should be apparent from the earlier discussions in this book that System Dynamics is based on identifying and describing the processes by which accumulations result in the world from rates of change. In mathematical terms it is based on calculus and levels are the integration of rates over a period of time. The amount accumulated in a level served by a single rate variable over a given time period is equal to the area under the graph of the rate plotted over that period of time.

It is the process of changing a rate into a level that creates a movement forward in time and understanding of this point is vital to understanding System Dynamics simulation. The laws for constructing System Dynamics diagrams, the time subscripts used in some System Dynamics software equation syntax and the computational sequence employed in System Dynamics simulation all depend on it.

In diagramming terms it follows that levels should depend on rates and rates only (never on auxiliaries or other levels), and that rates should depend only on information from auxiliaries and other levels (never on other rates). Level and auxiliary variables exist at a point in time and, thus, can be measured and used to determine rates, which are used to modify levels in subsequent time periods. It also follows that a feedback loop must contain at least one level and one rate; otherwise creation of behaviour *over time* from tracing out the loop would not be possible.

Rather than using analytical methods of solution from calculus, based on differential equations, System Dynamics employs numerical simulation methods based on simple difference equations to represent the process of accumulation.

Although simulation may involve hundreds of time points, it can be explained in terms of three points in time (see step N in Figure 4.1). In the syntax of DYSMAP2, K is defined as the current point in time, J as a past point in time and L as a future point in time; each separated from each other by a small interval of time DT, known as the simulation interval. An inherent assumption here is that DT should be sufficiently small relative to other time factors in the model to effectively represent an instantaneous time change and, hence, to approximate the flows to being continuous. Some guidance for determining the appropriate size of DT is given in Appendix 6.

In order to demonstrate the construction of simulation equations and their use, an example will be presented. This is the staff recruitment model shown in Figure 4.2 and represents a situation where it is required to recruit

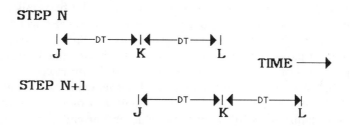

Figure 4.1   Time shift and relabelling in DYSMAP2 simulation

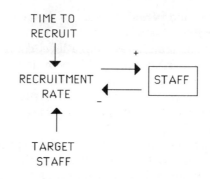

Figure 4.2 Influence diagram for the staff recruitment model

30 staff for a new department of a company. This can be considered either as a specific case of the general situation depicted in Figure 2.8a or as the other half of the staff leaving rate situation depicted in Figure 2.5.

It is important to recognise that what is required is to change a system state (staff level) from 0 to 30 and that a policy must be defined by which to achieve this. A simple and appropriate policy for this is that of proportional control encountered in Chapter 3, where information concerning the magnitude of the level (number of staff) is fed back to control the recruitment rate. The recruitment rate is calculated by dividing the discrepancy between the target and actual value of the level by a time factor, which represents the average delay in performing the recruitment.

The assumption is that recruits will be taken on as available and that the delay in recruitment will vary about the average due to variations in the time of response to the advertising of the post, availability of potential staff, the rejection rate and the notice potential staff require to resign from their previous positions. All these factors could be explicitly modelled, but are combined here for simplicity into an average recruitment delay.

SIMULATION EQUATIONS

The model of Figure 4.2 contains one level, one rate and two constants and, hence, requires equations to be specified for each of these variables. Since levels (and auxiliaries) are calculated at the current time they have a time subscript K. In particular, level equations always have a fixed format. They are simply based on the magnitude of the level at the previous point in time J, plus what has flowed into the level, and minus what has flowed out of the level, during the period JK. A level equation is denoted by the letter L (not to be confused with the time point L) and an initial condition or N equation must be specified for it which should not have any time subscript.

The equation and its initial condition for the level variable STAFF in Figure 4.2 is as follows:

L STAFF.K=STAFF.J+DT*(RECRUIT_RATE)   (4.1)

(people) (people) (month) (people/month)

N STAFF=0   (4.2)

Equation 4.1 also indicates the dimensions of each of its components. It is vital that great care is given to dimensioning variables to ensure that both sides of each equation are dimensionally consistent.

Rate equations are denoted by a letter R and the time subscript for the rate variable is KL, indicating a calculation for the future time period. The logic is that levels (and auxiliary variables) are calculated at the current time and rates projected for the forthcoming time period. Time is then moved forward in the simulation software by DT, hence, time point K becomes J, time point L becomes K and, consequently, the time period KL becomes JK. This allows recalculation of all levels and auxiliaries at the new time K and rates for the next period KL (see Step N+1 in Figure 4.1).

As already indicated, the equation for the rate variable RECRUIT_RATE in Figure 4.2 is based on proportional control.

R.RECRUIT_RATE.KL
        =(TARGET_STAFF.K-STAFF.K)/RECRUIT_DELAY   (4.3)

(people)   (people)   (months)

The intention is to eliminate a proportion of the discrepancy between the target and actual values of staff in each time period. This proportion is (1/RECRUIT_DELAY). If RECRUIT_DELAY was set to a value of 1 all the discrepancy would be cleared every month. As RECRUIT_DELAY increases the RECRUIT_RATE decreases in magnitude giving a smoother response to changes in the staff discrepancy.

Equations for constants are denoted by the letter C and the two constants in Figure 4.2 can be represented as follows (using values for target staff of 30 people and for recruitment delay of 5 months):

C TARGET_STAFF=30   (4.4)

C RECRUIT_DELAY=5   (4.5)

SYSTEM BEHAVIOUR

Having created a diagram relating the resource flows and information–behavioural flows in a system, and having created equations representing the relationships between the defined variables, the next step is to see what behaviour is produced by each variable over time. In other words to see what happens when the equations are processed to determine the values of each variable at the end of each future time period. This processing is known as simulation and the computational sequence for this is shown in Figure 4.3.

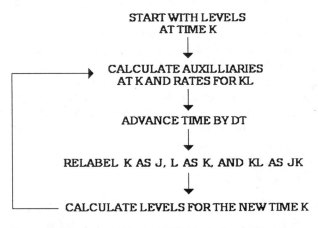

Figure 4.3 Computational sequence in DYSMAP2 simulation

Simulation can be performed either by hand or computer. Obviously, the latter is the ideal approach since the necessary 'number crunching' is very tedious and repetitive. However, hand simulation provides an excellent way by which to reinforce both the mechanism of System Dynamics simulation (and hence the time subscripts used in the DYSMAP2 computer simulation language) and the effects of policy change on system behaviour. These attributes of hand simulation will be demonstrated by simulating the Staff Recruitment Model.

HAND SIMULATION

Figure 4.4 gives a tabulated output from simulating equations 4.1 to 4.5 by hand on the assumption that DT is 1 week and the simulation length is 10 weeks. The generation of the output proceeds as follows:

Time	Staff	Target_Staff	Target_Staff-Staff	Recruit_Rate
0.0	0.0	30.0	30.0	6.0
1.0	6.0	30.0	24.0	4.8
2.0	10.8	30.0	19.2	3.8
3.0	14.6	30.0	15.3	3.0
4.0	17.7	30.0	12.3	2.4
5.0	20.2	30.0	9.8	2.0
6.0	22.1	30.0	7.9	1.6
7.0	23.7	30.0	6.3	1.3
8.0	25.0	30.0	5.0	1.0
9.0	26.0	30.0	4.0	0.8
10.0	26.8	30.0	3.2	0.7

Figure 4.4 Hand simulation output for the Staff Recruitment Model

(i) At time zero the current staff level is given by the initial condition of the staff level (zero) (equation 4.2).

(ii) The rate of recruitment to be applied in the next time interval is calculated from the current value of STAFF, TARGET_STAFF AND RECRUIT_DELAY using the RECRUIT_RATE equation (equation 4.1).

(iii) Time is now moved forward 1 DT and the rate of recruitment from (ii) implemented using the level equation for STAFF (equation 4.3).

(iv) Steps (ii) and (iii) are then repeated until the required length of the simulation is reached.

The procedure demonstrates the logic of the simulation sequence, that is, rates can only depend on current information about levels, constants and auxiliaries and that levels and levels only are changed by rates.

COMPUTER SIMULATION

A full set of DYSMAP2 simulations equations for the Staff Recruitment Model is shown in Figure 4.5. To perform simulation in practice other statements are necessary in addition to those already defined. These are to specify the length of simulation, the simulation interval, the printing interval and a run statement. In fact, two run statements are included in Figure 4.5. The second after modifying the value of DT so that the effect of this change can be tested. A full documentation of all variables and their dimensions is also given. The documentation equations are denoted by the

```
NOTE            FIGURE 4.5          (MODEL STAFFB)
NOTE
NOTE            DYSMAP2 EQUATIONS FOR THE STAFF
NOTE            RECRUITMENT MODEL USED TO CREATE
NOTE            THE HAND SIMULATION RESULTS OF FIGURE 4.4
NOTE      AND THE COMPUTER SIMULATION RESULTS OF FIGURES 4.6-4.9
NOTE
L STAFF.K=STAFF.J+DT*(RECRUIT_RATE.JK)
N STAFF=INITIAL_STAFF
C INITIAL_STAFF=0
R RECRUIT_RATE.KL=(TARGET_STAFF-STAFF.K)/RECRUIT_DELAY
C RECRUIT_DELAY=5
C TARGET_STAFF=30
SPEC LENGTH=25/DT=1/PRTPER=1
PRINT 1)STAFF
N TIME=0
RUN    FIGURE 4.7 RECRUITMENT MODEL WITH DT=1
C DT=0.1
RUN RECRUITMENT MODEL WITH DT=0.1
PRINT 1)STAFF
PRINT 2)TARGET_STAFF
NOTE
NOTE    DOCUMENTATION OF VARIABLES
NOTE
D STAFF=(PEOPLE)    NUMBER OF CURRENT STAFF MEMBERS
D RECRUIT_DELAY=(MONTH)      AVERAGE TIME TO RECRUIT STAFF
D RECRUIT_RATE=(PEOPLE/MONTH) RATE OF RECRUITMENT OF STAFF
D TARGET_STAFF=(PEOPLE)   TARGET NUMBER OF STAFF
D TIME=(MONTH)    SIMULATION TIME
D LENGTH=(MONTH)   SIMULATION LENGTH
D PRTPER=(MONTH)   SIMULATION INTERVAL
D DT=(MONTH)   SIMULATION INTERVAL
D INITIAL_STAFF=(PEOPLE)   INITIAL NUMBER OF STAFF
```

Figure 4.5 DYSMAP2 equations for the staff recruitment model used to create the hand simulation results of Figure 4.4 and computer simulation results of Figures 4.6 to 4.9

letter D and DYSMAP2 uses these to perform dimensional analysis of all the equations.

Figures 4.6 and 4.7[2] show numerical and graphical computer simulation output, respectively, of the results of simulating the Staff Recruitment Model over 25 time periods using the DYSMAP2 software. Figure 4.6 will be seen to correspond exactly to the results in Figure 4.4. The advantage of the graphical interpretation of the output in Figure 4.7 will be appreciated. This shows how the staff level rises smoothly to the target value.

Recruitment Model		Recruitment Model	
Time	Model	Time	Staff
0.0000	0.0000	14.000	28.681
1.0000	6.0000	15.000	28.944
2.0000	10.800	16.000	29.156
3.0000	14.640	17.000	29.324
4.0000	17.712	18.000	29.460
5.0000	20.170	19.000	29.568
6.0000	22.136	20.000	29.654
7.0000	23.709	21.000	29.723
8.0000	24.967	22.000	29.779
9.0000	25.973	23.000	29.823
10.0000	26.779	24.000	29.858
11.0000	27.423	25.000	29.887
12.0000	27.938		
13.0000	28.351		

Figure 4.6 Numerical output from computer simulation of the Recruitment Model with DT=1

Figure 4.8 shows the numerical results from a repeat of the experiment with a DT of 0.1, but with the output still printed out every month. The results arise from the second RUN statement in the equation listing in Figure 4.5. The value of DT used should be much smaller than the time units of the model and the results of Figure 4.8 are the more accurate, there being ten calculations every month rather than one. A value of 1 month was used in the hand simulation to simplify the calculations. In practice there is a trade off between using a very small value of DT for improved accuracy and the additional calculations and, hence, simulation run time incurred.

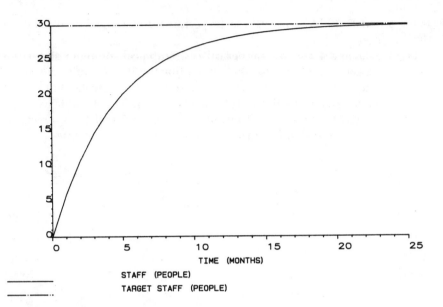

Figure 4.7 Recruitment Model with DT=1

Recruitment Model		Recruitment Model	
Time	Staff	Time	Staff
0.0000	0.000	14.000	28.227
1.0000	5.487	15.000	28.551
2.0000	9.971	16.000	29.816
3.0000	13.635	17.000	29.033
4.0000	16.629	18.000	29.210
5.0000	19.075	19.000	29.354
6.0000	21.073	20.000	29.472
7.0000	22.706	21.000	29.569
8.0000	24.041	22.000	29.648
9.0000	25.131	23.000	29.712
10.0000	26.021	24.000	29.765
11.0000	26.749		
12.0000	27.344		
13.0000	27.830		

Figure 4.8 Numerical output fom computer simulation of the Recruitment Model with DT=0.1

Figure 4.9 shows a graphical comparison of the results from using the different values of DT. (This presentation makes use of the CO-PLOT facility in DYSMAP2, whereby the same variable from different computer runs of a model can be superimposed on one another.)

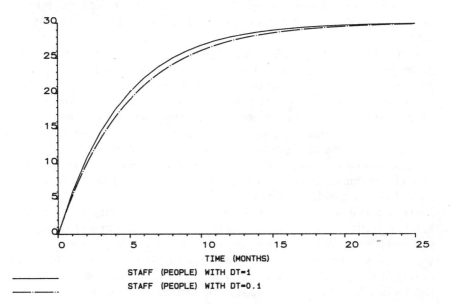

Figure 4.9 Staff Recruitment Model with different values of DT

It should be pointed out that the values of DT used here are purely for demonstration purposes. In practice, it is recommended that DT be specified as a binary function (that is, one which is divisible by 2), to avoid possible instabilities in more complex models.

SOFTWARE

It is important to emphasise that the simulation of System Dynamics models can be carried out in any computer language. Obviously, high level computer languages are most appropriate and better still is the use of purpose built software, since these encompass algorithms which facilitate the construction of dynamic models. For example, they contain sort routines to place simulation equations into a computable sequence and include a host of functions to assist with the formulation of specific model relationships and policies.

The original System Dynamics software was DYNAMO and, in addition to the version for use on mainframe computers, this is now available as Professional DYNAMO (Pugh Roberts Associates 1986) for use on the IBM PC range of computers. DYSMAP2 (Dangerfield and Vapenikova 1987), as used in parts of this book, uses a similar equation format to Professional DYNAMO and is specifically designed for use on the IBM PC range of computers. Both of these programs facilitate rigorous development of large models.

In addition, other purpose built System Dynamics software has been developed in recent years. The most recent and perhaps significant development, has been that of STELLA (Richmond *et al.* 1987). This is for the Apple Macintosh range of computers and takes full advantage of the hard-wired graphics interface of these computers. Its major innovation is that pipe diagrams representing models can be drawn directly on the computer screen using a predefined tool kit. The variables of the diagrams are represented by icons which can be opened to insert parameter values and relationships between variables. The diagrams in Figures 2.2a and 2.3a were drawn using this software.

STELLA is a particularly useful and appealing language for fast model building by those users who are less familiar with computer programming. It is excellent for demonstrating the relationship between system feedback structures and system behaviour and for involving system actors more closely in the model building and analysis process. This software is being rapidly developed to provide ease of interaction of simulation models with other software, such as Hypercard, (Richmond and Peterson 1988) and to help modellers create purpose built user interfaces for data input and output.

Examples of both DYSMAP2 and STELLA type models will be found in later chapters.

VALIDATION

Once results are obtained from a simulation model, the question arises as to whether it is a valid representation of the real world system which was modelled.

In many fields of enquiry the validity of a model often refers only to whether it can accurately reproduce past statistical data as observed in the real system. Although it is considered important in System Dynamics that a model can reproduce a reference mode of behaviour, this is seen as only one of a range of tests of validity. Exact correspondence between a model output and past data is perhaps significant for certain types of models (for example, econometric models) which are used for absolute, short term forecasting. For System Dynamics models, which attempt to create scenarios based on assumptions about multiple relationships and policies (which might be totally different from those of the past), such emphasis on the past is seen as less important.

In System Dynamics models validity is seen as a more complex concept which centres on user confidence in the model. This confidence stems from an appreciation of the structure of the model, its general behaviour characteristics and its ability to generate accepted responses to set policy changes.

The summary of tests for building confidence in System Dynamics models is given in Table 4.1, which is adapted from Sterman (1984).

In the case of the simple model considered in this chapter, validity is achieved in the sense that it structurally represents the reality described and produces an accepted response to a given input, that is, a smooth transition when subjected to a proportional control policy.

IMPROVEMENT OF SYSTEM BEHAVIOUR

The most important role of a System Dynamics simulation model is to facilitate experimentation and, hence, to design system structures and strategies for improved system behaviour.

The whole process of system design can be considered as a cyclic procedure as shown in Figure 4.10. This figure demonstrates how alternative parameter values, structures and strategies are tested in simulation experiments to create model outcomes in terms of numerical and graphical behaviour of system variables over time (or specific performance measures). These, in turn, lead to improved understanding and, hence, to further changes of parameter values, structure and strategy.

Figure 4.10 also allows for the use of an external driving force in creating experiments with a model.

This inclusion captures a fundamental debate in System Dynamics, as to whether a model should be totally endogenous (closed) or allowed to have exogenous (external inputs). The American tradition is to build the former type of model to emphasise the need for models to be endogenously capable of producing behaviour observed in the real world. The more European approach is to use external inputs or driving forces to disturb models from equilibrium and to test the robustness of strategies and policies under a range of external influences. This type of approach is very compatible with control theory practice, although it should be stressed that exogenous factors should be restricted to a very small number. The decision between partially open or fully closed models depends on the system under the study and both types of model will be found in this book.

In the case of exogenously driven models, the driving force can be the

Table 4.1 Tests for building confidence in System Dynamics models*

Test of model structure	Question addressed by the test
Structure Verification	Is the model structure consistent with relevant descriptive knowledge of the system?
Parameter Verification	Are the parameters consistent with relevant descriptive (and numerical, when available) knowledge of the system?
Extreme Conditions	Does each equation make sense even when its inputs take on extreme values?
Boundary Adequacy (Structure)	Are the important concepts for addressing the problem endogenous to the model?
Dimensional Consistency	Is each equation dimensionally consistent without the use of parameters having no real-world counterparts?

Tests of model behaviour	
Behaviour Reproduction	Does the model generate the symptoms of the problem, behaviour modes, phasing, frequencies and other characteristics of the behaviour of the real system?
Behaviour Anomaly	Does anomalous behaviour arise if an assumption of the model is deleted?
Family Member	Can the model reproduce the behaviour of other examples of systems in the same class as the model (e.g. can an urban model generate the behaviour of New York, Dallas, Carson City and Calcutta when parametrised for each)?
Surprise Behaviour	Does the model point to the existence of a previously unrecognised mode of behaviour in the real system?
Extreme Policy	Does the model behave properly when subjected to extreme policies or test inputs?
Boundary Adequacy (Behaviour)	Is the behaviour of the model sensitive to the addition or alteration of structure to represent plausible alternative theories?
Behaviour Sensitivity	Is the behaviour of the model sensitive to plausible variations in parameters?
Statistical Character	Does the output of the model have the same statistical character as the 'output' of the real system?

Tests of policy implications	
System Improvement	Is the performance of the real system improved through use of the model?
Behaviour Prediction	Does the model correctly describe the results of a new policy?
Boundary Adequacy (Policy)	Are the policy recommendations sensitive to the addition or alteration of structure to represent plausible alternative theories?
Policy Sensitivity	Are the policy recommendations sensitive to plausible variations in parameters?

*Reprinted from Sterman 1984. Reproduced with the permission of DYNAMICA.

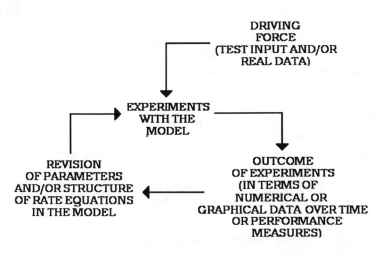

Figure 4.10 The process of policy design in System Dynamics

input of real data or a synthetic shock, generated by a test input such as a step function or sine wave. The idea of the use of test inputs is to create a controlled situation against which to assess the capability of alternative structures and strategies in moving the system through a transitional phase.

Irrespective of the approach, there are a number of concepts on which changes to structure and strategies can be based:

Intuition of Systems Actors and Owners

Alternative structures and strategies for testing can usually be generated by people with experience of the system under study. Many of these have their own 'pet' theories as to improvements which can be made. This is a good and easily accessed source of ideas.

Use of Control Engineering Concepts for Designing System Parameters

A System Dynamics model is analogous to a control engineering model and, hence, it is prudent to look to this field to provide some design theories. In control engineering systems, structures are very fixed and the scope for changing control comes mainly from changing system *parameters* used within strategies. Hence, design in control engineering primarily consists of parameter tuning.

In simple control engineering systems (with restricted numbers of states or levels), and even in more complex systems involving non-linear components, assumptions can be made to apply such parameter tuning by analytical methods.

In contrast, most socio-economic and management systems are sufficiently complex and non-linear to make direct parameter tuning approaches difficult to apply. There is, however, much research along these lines involving the use of such concepts as eigenvalue analysis and optimal control theory (Mohapatra 1980, Sharma 1985). Rules of thumb have also been derived from control theory results for parameter policy design in socio-economic systems (Coyle 1977).

A further area of parameter policy design under current research is that associated with analysis of system parameters for which system behaviour is unstable. In particular, the role of deterministic chaos is being explored (Mosekilde et al. 1988) as a means of predicting such regions.

The Design of System Structure and Strategy

Unlike control engineering systems, the process, information and organisation structures and strategies of socio-economic systems are totally flexible and enormous scope exists by which to redesign them. The greatest flexibility for improved design lies in the information structure of systems and it is this which is usually referred to when structural changes are considered in System Dynamics models. The information structure comprises the actual information links feeding strategy variables, hence, changing information structure means changing information on which the strategies are based, rather than (as in parameter policy design) just changing the parameters of the strategy itself.

Guidance for this design procedure has already been discussed for Qualitative System Dynamics in Chapter 3. Extensive simulation experimentation with a model allows such ideas to be comprehensively explored and, to a large extent, the analysis becomes self perpetuating. The understanding generated from one set of results often suggests the direction for the next experiment.

Use of Heuristic Optimisation

One of the more recent methods developed for policy design in System Dynamics models is that of heuristic optimisation (Keloharju 1983, Wolstenholme and Al-Alusi 1987). The procedure involves linking a heuristic optimisation algorithm to a System Dynamics model and using this to choose parameters values, within given ranges, to optimise model performance as specified by particular objective functions (performance measures). This is a powerful method which, although working on system parameters, can by suitable creation of pseudo model parameters, also be applied to information structure design. This method is considered in detail in Chapters 9 and 10.

SUMMARY

This chapter has introduced the basic concepts and philosophy of model quantification in System Dynamics. The superimposition of numbers on model parameters and the creation of equations to represent relationships between variables, adds a new dimension to modelling. It also permits the power of the computer to be used to simulate the behaviour of the model over time.

As indicated in Chapter 1, the use of a computer requires a learning and familiarisation process. However, once the format of equations for a particular piece of simulation software have been assimilated, modellers have at their disposal a tool to facilitate extensive experimentation.

Further components of Quantitative System Dynamics modelling will be introduced in the next chapter.

NOTES

[1] The version of DYSMAP2 used throughout this book is the basic version for use on IBM or compatible P.C.'s incorporating 286 chips. An enhanced version of the software known as DYSMAP2/236 is also available (Vapenikova, 1989). This version allows significantly faster processing and the capability to run extremely large models. This enhanced performance is achieved through the use of a DOS extender called DBOS, a piece of software which permits the 386 chip to operate in real mode with an addressable memory of 2 gigabytes. However, to the user, the functionality of DYSMAP2 and DYSMAP2/386 is identical.

[2] This and other DYSMAP2 graph-plots appearing in the book have been pen-plotted using the PCPLOT (Jackman 1988) software. PCPLOT is a separate graph-plotting program developed from the DYSMAP2 multiple axis graph-plot option contributed by Robert Jackman. It features high quality plotted output on Hewlett-Packard 7550 and compatible plotters in addition to the standard screen and printer facilities offered within DYSMAP2.

REFERENCES

Coyle, R.G. (1977) *Management System Dynamics*, Wiley, Chichester.

Dangerfield, B. and O. Vapenikova (1987) *DYSMAP2 User Manual*, University of Salford.

Jackman, R. (1988) *PCPLOT User Manual*, University of Salford.

Keloharju, R. (1983) *Relativity Dynamics* Helsinki School of Economics, Helsinki, Finland.

Mohapatra, P.K.J. (1980) *Part 1—Structural Equivalence Between Control System Theory and System Dynamics, Part II—Non-linearity in System Dynamics Models*, DYNAMICA vol. 6, pt. 1.

Mosekilde, E., J. Aracil and P.M. Allen (1988) Instabilities and Chaos in Non-Linear Dynamic Systems, *System Dynamics Review*, **4**, 56–81.

Pugh-Roberts Associates (1986) *Professional DYNAMO Introductory Guide and Tutorial and Professional DYNAMO Reference Manual*, Pugh-Roberts Associates, Five Lee Street, Cambridge, MA.

Richmond, B.M. and S. Peterson (1988) *A User's Guide to STELLAstack*, High Performance Systems Inc.

Richmond, B.M., P. Vescuso and S. Peterson (1987) *STELLA for Business*, High Performance Systems Publications, 13 Dartmouth College Highway, Lyme, NH.

Sharma, S.K. (1985) *Policy Design in System Dynamics Models: Some Control Theory Applications*, Doctoral Thesis submitted to the Indian Institute of Technology, Kharagpur, India.

Sterman, J.D. (1984) *Appropriate Summary Statistics for Evaluating the Historic Fit of System Dynamics Models*, DYNAMICA, vol. 10, pt. 11 winter 1984.

Vapenikova, O. (1989) *DYSMAP2/386—DYSMAP2 Implementation on 80386—Based P.C.'s*, Proceedings of the European Simulation Multi-Conference, Rome, 1989, published by Society for Computer Simulation International.

Wolstenholme E.F. and S.A. Al-Alusi (1987) System Dynamics and Heuristic Optimisation in Defence Analysis, *System Dynamics Review*, **3**, 102–116.

Chapter 5

Quantitative System Dynamics II— Equation Formulation, Model Development and Policy Analysis

INTRODUCTION

The purpose of this chapter is to give some concrete examples of *structural policy analysis* in System Dynamics. The Staff Recruitment (SR) Model created in Chapter 4 will be developed both for this purpose and to show how System Dynamics simulation equations can be constructed to represent other concepts. These are, in turn, the smoothing of information, non-linear functions, multipliers, and delays. The idea of equilibrium analysis will also be introduced.

The model development will be carried out in the DYSMAP2 language and introduce a range of functions available within this. A list of all DYSMAP2 functions used in this book is presented in Appendix 6 and a full list can be found in the DYSMAP2 User Manual (Dangerfield and Vapenikova 1987).

It is possible in DYSMAP2 to change *any* equation following a run statement and, hence, to create multiple runs or models within one listing of equations. In the original DYSMAP, and other software, such changes were restricted to constants equations only which made the construction of multiple models somewhat clumsy. (In fact, using DYSMAP2 it is possible by changing equations to create additional runs interactively.)

The use of the multiple model facility of DYSMAP2 has been used extensively in this chapter. However, the main purpose of multiple models is to allow comparison plotting (co-plots) of variables over all models. Since DYSMAP2 restricts the number of variables in any plot to 6 for the purpose of clarity of results on a computer monitor, it is considered redundant to subsume more than six models or runs within one equation listing, hence, the eleven models developed in this chapter have been grouped into two equation listings of five and six runs respectively. These equation listings are presented in Tables 1 and 2 in Appendix 1.

A STAFF RECRUITMENT AND LEAVING (SRL) MODEL

Chapter 4 introduced a simple single level (single order), composite System Dynamics model. This was the Staff Recruitment Model of Figures 4.2 and 4.5. Figure 5.1 shows an influence diagram of this model extended to incorporate a leaving rate. This new model will be referred to as the Staff Recruitment and Leaving (SRL) Model.

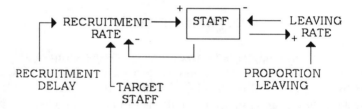

Figure 5.1 Influence diagram of Staff Recruitment and Leaving Model

Table 1 in Appendix 1 gives the DYSMAP2 equations for the new model. These are contained between the beginning of the model and the line stating 'RUN 1 SRL MODEL WITH PROPORTIONAL CONTROL'.

It will be seen that the level equation for STAFF now includes a LEAVING_RATE, which is defined as equal to the number of staff multiplied by a proportion leaving per month. This is assumed to be 5% per month. Note the units for this variable in the documentation section are given as (1/month). Also note that DT is set to 0.1 and the constants LENGTH, DT and PRTPER (printing period) are given in a SPEC (specification) statement.

As indicated by the title of the run, the policy used for RECRUIT_RATE was that of proportional control as described in Chapters 3 and 4.

Graphical output from the SRL model is shown in Figure 5.2. It will be noticed that unlike the original SR model, the staff variable in this new model does not attain its target value of 30, but only reaches a value of 24.

The reason for this situation is that the proportional recruitment policy which worked well for the SR model is inadequate for the SRL model and this statement emphasises the importance of the whole concept of policy design.

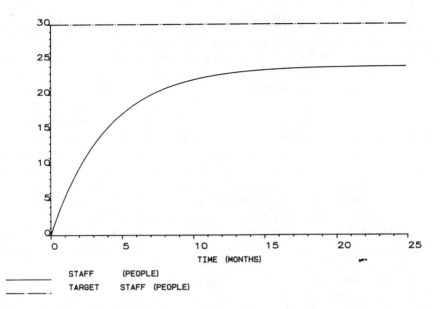

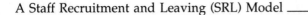

Figure 5.2 Plot of 'staff' and 'target staff' from SRL model with proportional control of recruitment rate

The SLR model will, in fact, reach equilibrium when the recruitment rate is equal to the leaving rate and, as shown below, using the equations for the model without time subscripts, this occurs at a staff level of 24:

At equilibrium

```
RECRUIT_RATE = LEAVING_RATE                                     (5.1)
```

i.e.

```
(TARGET_STAFF-STAFF)/RECRUIT_DELAY=
                        STAFF * PROPORTION_LEAVING (5.2)
```

i.e.

```
STAFF = TARGET_STAFF/
            (1+(RECRUIT_DELAY*PROPORTION_LEAVING))        (5.3)

     = 30 /(1+(5*0.05))

     = 30 /1.25

     = 24
```

Given the inadequacy of the proportional recruitment rate policy in this case, it is necessary to experiment with alternative policies. The remainder of the Chapter is concerned with a range of such experiments, some of which will require the introduction of new concepts.

Runs 2 and 3 in Table 1, Appendix 1 contain changes to the equations of the basic SRL model for *parameter* based policy experiments, involving different values of the recruitment delay, in order to try to overcome the problem, however, as shown in Figure 5.3 changing the recruitment delay does not make any improvement. With a recruitment delay of 2 months, the staff level reaches a value of around 27 and with the recruitment delay set to 10, the staff level only reaches a value of 20. In other words parameter policy design does not solve the problem of the staff level not reaching its target. The equilibrium staff level achieved for the different values of the recruitment delay parameter can be verified by repeating the previous manual calculations of equations 5.1 to 5.3 for these values.

A little thought will lead to the realisation that the policy of proportional control worked fine until the *leaving rate* was introduced into the model. Hence, adding the *leaving rate* to the recruitment rate equation might improve things. However, the leaving rate is, by definition a rate variable which, as described in Chapter 2, cannot be measured unless averaged, thus prior to testing out the new policy (which, incidentally, is a *structural* policy

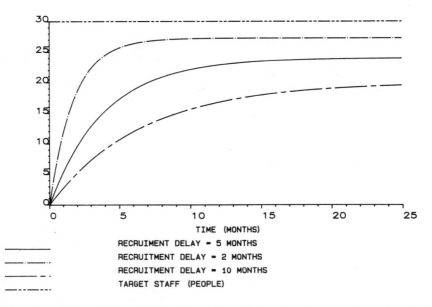

TIME (MONTHS)

RECRUIMENT DELAY = 5 MONTHS
RECRUITMENT DELAY = 2 MONTHS
RECRUITMENT DELAY = 10 MONTHS
TARGET STAFF (PEOPLE)

Figure 5.3 Plot of 'staff' and 'target staff' from SRL model with proportional control of recruitment rate, using different recruitment delays

change) the DYSMAP2 equations for smoothing or averaging of information must be introduced.

THE AVERAGED LEVEL

As indicated in Chapter 2, the most commonly used approach for averaging information in System Dynamics models is to use exponential smoothing, where a weighted average is created between past data, as reflected in the old value of the average, and new data. The basic equation for exponential smoothing is

$$A_t = \alpha C_t + (1 - \alpha)A_{t-1} \tag{5.4}$$

where A_t is the average at time t, A_{t-1} the average at time $t - 1$, α the weighting factor and C_t the current value at time t.

Equation 5.4 can be rearranged as follows

$$A_t = A_{t-1} + \alpha (C_t - A_{t-1}) \tag{5.5}$$

This is the same form as the resource level equation and in DYSMAP2 syntax can be rewritten as follows:

```
L  A.K.=A.J.+(DT/SM) * (C.JK-A.J)
```
(5.6)
```
N  A=C
```

where α is replaced by $1/SM$ and SM is the time over which the average is made. Note that the dimensions of this type of level are units/time and that the the usual way to create an initial condition is by equating the average to the current value.

THE SRL MODEL WITH LEAVING RATE CONTROL

Run 4 in Table 1, Appendix 1 gives a listing of the equations of the SRL model with proportional plus inertial control. That is, the leaving rate is exponentially averaged over three months and added back to the original recruitment rate equation. The average leaving rate is referred to as an inertial component of the system.

Figure 5.4 shows a co-plot of the staff variable over time from both policies considered so far and it will be appreciated that inclusion of the inertia term enables the staff target to be achieved.

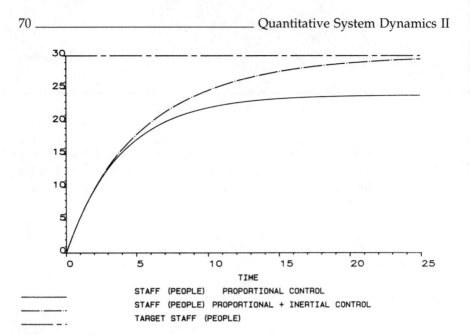

STAFF (PEOPLE)   PROPORTIONAL CONTROL
STAFF (PEOPLE) PROPORTIONAL + INERTIAL CONTROL
TARGET STAFF (PEOPLE)

Figure 5.4 Plot of 'staff' from SRL model with proportional control and proportional plus inertial control of recruitment rate

THE SRL MODEL WITH FIXED RECRUITMENT RATE

It is of interest to compare the merits of the above policies with a more conventional approach of assuming a fixed recruitment rate policy (of say 5 people/month) and simply switching off recruitment when the target is achieved.

Run 5 in Table 1 in Appendix 1 gives the equations for a DYSMAP2 model to achieve this and Figure 5.5 the graphical output. The staff level does indeed attain the target value but, of course, the recruitment rate has to be alternatively switched on and off to maintain it. Whilst, this may be acceptable in this system, there are many cases where it is not. For example, it would not be possible in a production system to have a stop–start production rate, also, the simple model here considers only one transition to a new value of target staff. In practice the target level itself might behave dynamically as plans change. Considerable instabilities would result from continually trying to adjust to such changes by discrete changes in the controlling rate.

In general, a policy of smooth response is a characteristic of the System Dynamics approach to change and this is basically an emulation of nature. Natural responses are predominantly smooth and the idea of instant action can be considered as a human concept.

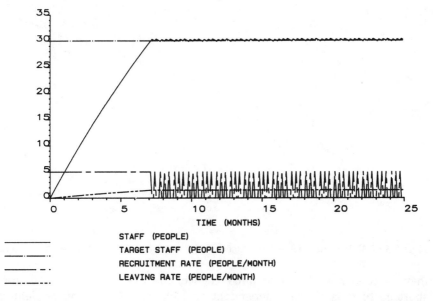

Figure 5.5 Plot of 'staff', 'target staff', 'recruitment rate' and 'leaving rate' for SRL model with a fixed recruitment rate

THE SRL MODEL WITH NONLINEAR CONTROL

Rate or policy equations in system dynamics models can take any form and it is possible to construct and test specific non-linear policies to describe the way in which nature or managers operate. This facility is made possible in most system dynamics software by the use of TABLE functions.

There are many different forms of this function in DYSMAP2, but the one predominantly used is the TABHL function (see Appendix 6). Figure 5.6 gives a recruitment rate policy which will now be incorporated into the SRL model. The table is a plot of the staff recruitment rate (dependent variable) against the discrepancy between target staff and staff (independent variable). The table presents a policy that sets the recruitment rate to zero when the discrepancy is zero, but increases the recruitment rate as the discrepancy increases. It reaches 5 per month at a staff discrepancy of 30 people.

Run 6 in Table 2, Appendix 1 shows the DYSMAP2 equations for the SRL model incorporating the policy described in Figure 5.6. The recruitment rate is a TABHL function. Such a table puts a horizontal tail onto the curve at the heights given by the last and first values in the table, and uses these for the result if the independent variable is respectively greater

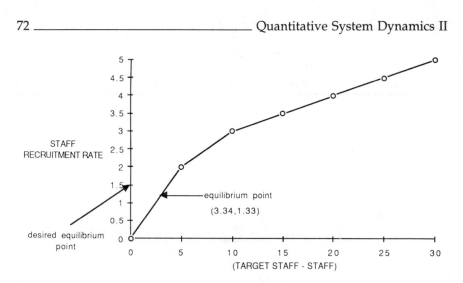

Figure 5.6 Policy table (POLTAB) used in the staff recruitment and leaving model

than or less than the table range, hence, the model will not fail from shortage of data if the independent variables exceed the range defined in the table. In the RECRUIT_RATE equation in Table 2, Appendix 1 the table is named POLTAB (for policy table) and the independent variable is defined as TARGET_STAFF-STAFF, together with its high (HI) and low (LO) values and increment (INC), all of which are constants. The corresponding dependent variable values are then defined in a T statement for POLTAB and intermediate values are interpolated.

Figure 5.7 shows the results of using this policy and it will be noted that, again, the system fails to meet its target. The equilibrium value of the staff level is, in fact, 26.66.

This situation can be explained in terms of the policy table used and emphasises the care needed in constructing such tables.

As before, the equilibrium situation for this model is at the point where the recruitment rate is equal to the leaving rate. When the staff is at the desired level of 30 the leaving rate is 1.5 people/month. Hence, the *desired* value of the recruitment rate to achieve a target staff level of 30 per month is 1.5 people per month.

The figure of 1.5 gives a rough order of magnitude for the calculation of the *actual* equilibrium rate achieved in the current model. The *actual* equilibrium recruitment rate is accurately determined by interpolation from the policy table in Figure 5.6 between values of recruitment rate of 0 and 2 and staff discrepancy of 0 and 5. Its value is 2/5 * (TARGET_STAFF-STAFF). Hence:

2/5 * (TARGET_STAFF-STAFF)
         = TARGET_STAFF*PROPORTION_LEAVING   (5.7)

numerically

```
2/5 * (30-STAFF) = TARGET_STAFF * 0.05
```

Hence

```
STAFF = 26.66
```

At this point

```
RECRUIT_RATE=2/5+(30-26.66)=1.33
```

and

```
TARGET_STAFF-STAFF=3.34
```

It will be seen from Figure 5.7 that the equilibrium recruitment and leaving rates are, indeed, 1.33 people per month.

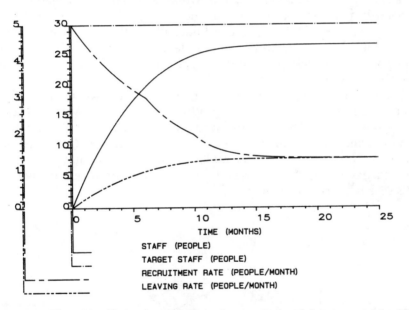

Figure 5.7 Plot of 'staff', 'target staff', 'recruitment rate' and 'leaving rate' for SRL model with non-linear control of recruitment rate

The problem arises because the policy table is not set up with equilibrium in mind. To achieve the target level of 30 staff the recruitment and leaving rates must be 1.5 people/month, hence, the policy table must give this value for the recruitment rate when the staff level equals its target value, that is, when the discrepancy is zero. This desired equilibrium point is marked on Figure 5.6. (As currently set up the policy table gives a recruitment rate of zero for a zero discrepancy.) Figure 5.6 also shows the actual equilibrium point for the model.

The policy table could be adjusted to achieve the desired equilibrium, however, defining a specific, numerical equilibrium point is only relevant as long as the values of the parameters in the model remain the same. For example, a new equilibrium point would need to be calculated for a different value of TARGET_STAFF. A much more comprehensive change would be to construct a policy table in a different way as explained in the next section.

THE SRL MODEL WITH MULTIPLIER CONTROL

The concept of multipliers is one of working in terms of deviations from a defined 'norm' point of a system and it is appropriate to establish this 'norm' as the equilibrium point. Additionally, if the multiplier is defined as a table function, where the independent axis is defined as a ratio, the function becomes independent of the specific values used for the variable on this axis.

Figure 5.8 describes a multiplier policy for the SRL model and the equations of run 7 in Table 2, Appendix 1 gives the DYSMAP2 equations incorporating this. The recruitment rate is now defined as the product of a normal recruitment rate (NOR_RECRUIT_RATE) equal to TARGET_STAFF* PROPORTION_LEAVING and the multiplier, hence, if the multiplier is equal to 1, the recruitment rate is equal to the normal recruitment rate, which in this case is 1.5 people/month.

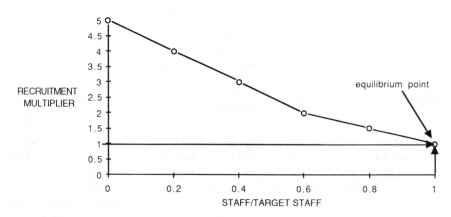

Figure 5.8 Multiplier policy function for staff recruitment and leaving model

It will be seen from Figure 5.8, that when STAFF = TARGET_STAFF the multiplier is defined as equal to 1. When STAFF are less than TARGET_STAFF (ratio < 1) the rate can increase up to 5 times above the normal level. (In this model the situation does not occur when STAFF exceeds TARGET_STAFF and hence no values of the multiplier are given in excess of 1, however, these could be defined, if appropriate.) The multiplier policy function is incorporated into the model using a TABHL function.

Figure 5.9 shows the results of using the multiplier table policy and it will be noted that the system now reaches its target value in a similar manner to that achieved using a combination of proportional and inertial control.

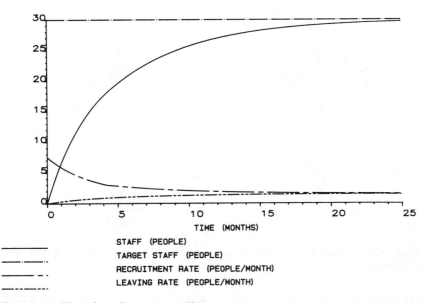

Figure 5.9 Plot of 'staff', 'target staff', 'recruitment rate' and 'leaving rate' for SRL model with multiplier control of recruitment rate

A FULLY BALANCED SRL MODEL

Rather than starting a system up from empty, it is better in practice to start it off in equilibrium and to disturb it from this by a test input (shock). The way in which the system reattains equilibrium can then be studied and the absence of dynamics prior to the test input can give much needed confidence that it is the test input, and only this, which is responsible for the post-shock dynamics. That is, there are no 'false' dynamics.

Run 8 in Table 2, Appendix 1 shows the equations for a fully balanced SRL Model using the multiplier policy for the staff recruitment rate. Figure 5.10 gives an influence diagram of this model. The value of STAFF is initially made equal to the value of TARGET_STAFF, that is, 30. Hence, prior to the step input the leaving rate has a value of 1.5 people per month. The staff ratio and the recruitment rate multiplier are both equal to 1, hence the recruitment rate is also equal to the normal recruitment rate of 1.5 people per month.

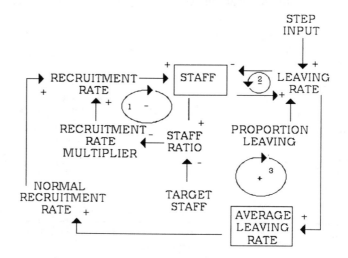

Figure 5.10 Influence diagram of staff recruitment and leaving model with multiplier control

The leaving rate is formulated as before except that a step input is included of height 5 at time 2. This will cause the leaving rate to move from its equilibrium value of 1.5 to 6.5 at time 5 months.

An average leaving rate is included which is used to define the normal recruitment rate. This enables the recruitment rate to track the leaving rate and allows the system to attain a new equilibrium level after the step input takes place.

Figure 5.11 shows a graphical output of the results of this model. It will be seen that the step in leaving rate at time 5 causes the staff level to fall. The recruitment rate then rises to compensate for this. However, because it is dependent on the normal recruitment rate which, in turn, is slightly delayed in its response by being dependent on the average leaving rate, it rises somewhat more slowly than the leaving rate. It actually continues on (or overshoots) after the leaving rate stops rising, but does eventually

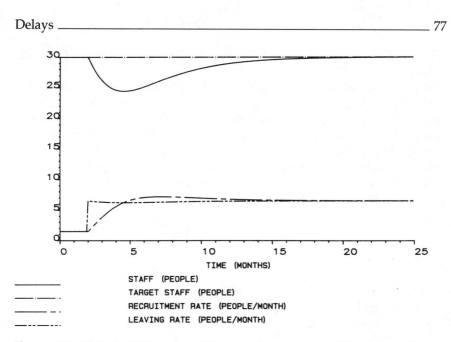

Figure 5.11 Plot of 'staff', 'target staff', 'recruitment rate' and 'leaving rate', for a fully balanced SRL model

reach its new equilibrium value (6.5 people/month) as does the staff level (30 staff).

It can be seen from Figure 5.10 that apart from the main negative control loop (1) by which the recruitment rate is controlled, the system has a further negative loop (2) linking staff and leaving rate and a positive loop (3) has been created linking recruitment rate, staff, leaving rate, average leaving rate and normal recruitment rate. The latter loop is responsible for delaying the effectiveness of the main negative control loop. The delay stems from the averaging process used within the loop. Delays are a major feature of dynamic behaviour and will be considered comprehensively in the next section.

DELAYS

The idea that delays can exist in both resource and information flows was introduced in Chapter 2. In both cases the concept of a delay implies that resources or information will be held up relative to other flows in the system. This suggests that delays can be represented by levels.

Figure 5.12 shows the simplest form of a *resource* delay in influence diagram form together with its DYSMAP2 equations. This delay, which consists of one delaying level only, is referred to as a first-order resource delay and has already been encountered in the basic structure of the SRL model.

Figure 5.13 shows a third-order *resource* delay and its associated equations, where the total delay (DEL) is split into three equal parts to control the resource flow between three levels. Delays of any order can be constructed by expanding the equations given in Figures 5.12 and 5.13.

In the DYSMAP2 software, macros exist by which to create the different orders of delays. For example, the set of equations in Figures 5.12 and 5.13 could be created, as shown, by means of DELAY1(INRATE,DEL) and DELAY3(INRATE,DEL) respectively.

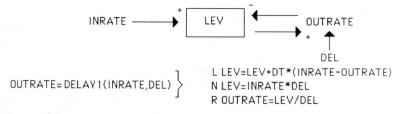

OUTRATE=DELAY1(INRATE,DEL)

L LEV=LEV+DT*(INRATE-OUTRATE)
N LEV=INRATE*DEL
R OUTRATE=LEV/DEL

Figure 5.12 Influence diagram, DYSMAP2 equations and macro for a first-order resource delay (Postscripts are omitted)

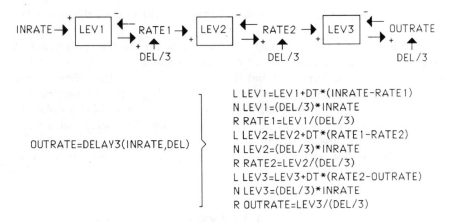

OUTRATE=DELAY3(INRATE,DEL)

L LEV1=LEV1+DT*(INRATE-RATE1)
N LEV1=(DEL/3)*INRATE
R RATE1=LEV1/(DEL/3)
L LEV2=LEV2+DT*(RATE1-RATE2)
N LEV2=(DEL/3)*INRATE
R RATE2=LEV2/(DEL/3)
L LEV3=LEV3+DT*(RATE2-OUTRATE)
N LEV3=(DEL/3)*INRATE
R OUTRATE=LEV3/(DEL/3)

Figure 5.13 Influence diagram, DYSMAP2 equations and macro for a third-order resource delay (Postscripts are omitted)

The concept of delays applied to *information* flows creates the idea of a level within an information flow. Such an information level has already been encountered in the SRL model and is commonly known as a smoothed level. This differs from a resource level and its name arises from the fact that *delaying* of information effectively represents *smoothing* of information. Figure 5.14 shows a first-order *information* level in influence diagram form with the associated equations. Figure 5.15 shows the same information for a third-order information delay. Again DYSMAP2 provides macros to represent information delays of different orders and the macros for the equations of Figures 5.14 and 5.15 are, as shown, DLINF1(INRATE,DEL) and DLINF3(INRATE,DEL), respectively. Note that the level equation representing the first-order information delay is exactly the same equation as that for the averaged information level introduced earlier in this chapter.

The easiest way to understand the effect of delays on system behaviour is to consider the response of a delay to a pulse input. Table 3 in Appendix 1 lists a model to demonstrate the output from a first-, third- and sixth-order resource delay, where the delay is 10 months. The latter is created by cascading together two third-order delays. The pulse function introduces a pulse of height HT at time ST and at an interval of INT months.

The output from this model is shown in Figure 5.16 where it will be seen that as the order of the delay increases the distribution of the output changes by becoming narrower and higher. Ultimately, if an infinite order

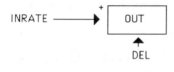

$$OUT = DLINF1(INRATE,DEL) \brace OR \ OUT=SMOOTH(INRATE,DEL)$$

L OUT= OUT + DT /DEL *(INRATE−OUT)
N LEV=INRATE

Figure 5.14 Influence diagram, DYSMAP2 equations and macro for a first-order information delay (Postcripts are omitted)

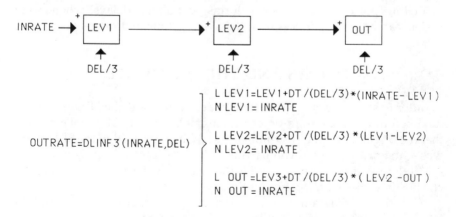

OUTRATE=DLINF3(INRATE,DEL)

L LEV1=LEV1+DT/(DEL/3)*(INRATE− LEV1)
N LEV1= INRATE

L LEV2=LEV2+DT /(DEL/3) *(LEV1−LEV2)
N LEV2= INRATE

L OUT =LEV3+DT /(DEL/3) * (LEV2 −OUT)
N OUT = INRATE

Figure 5.15 Influence diagram, DYSMAP2 equations and macro for a third-order information delay (Postcripts are omitted)

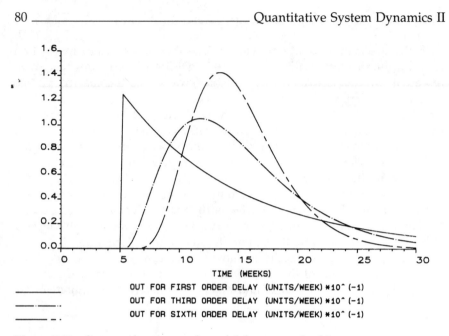

Figure 5.16   Output of various orders of delay to a pulsed input

delay was used (which can be created in DYSMAP2), the output would be a pulse output, exactly like the input, occurring exactly DEL units of time later.

The choice of the order of delay depends on the modelling context but a third-order delay is commonly used to create an approximately normal distribution of output from a given input. This is very appropriate to modelling process such as some forms of training delays, where some people can be trained in less than the average time, but others take more.

DELAYS AND THE SRL MODEL

In order to show the effect of delays on policy analysis, the SRL model was modified to include a third-order training delay. It should now be realised that the original training (recruitment) delay incorporated in the SRL model was a first-order delay. A comparison of Figures 5.1 and 5.12 should clarify this. (Note that in Figure 5.1 both the recruitment and leaving rates are effectively subjected to first-order delays.)

The influence diagram for the new model is given in Figure 5.17 and the DYSMAP2 equations in Run 9 in Table 2, Appendix 1.

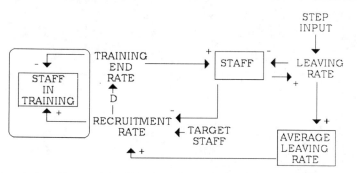

Figure 5.17 Influence diagram of staff recruitment and leaving model with a third-order training delay

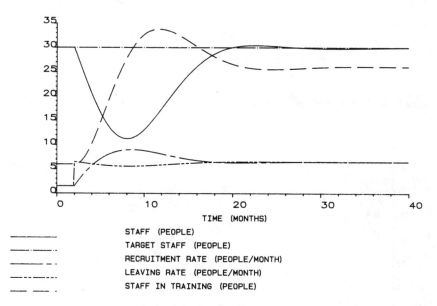

Figure 5.18 SRL model with third-order training delay and proportional plus inertial control

The recruitment rate control used was that of proportional plus inertial control as encountered in Run 4.

Figure 5.18 shows the response of the model to a step input in the leaving rate. The presence of the training delay causes some oscillation before a new equilibrium is reached and it was found necessary to run the model for at least 35 months to achieve stability. This is as expected from the ideal mode of behaviour of a delayed negative loop shown in Figure 2.9.

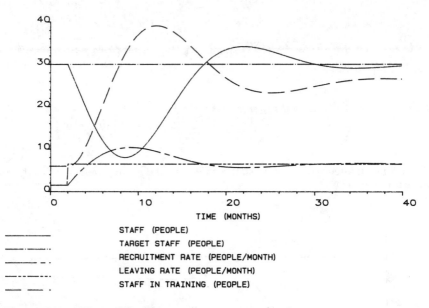

STAFF (PEOPLE)
TARGET STAFF (PEOPLE)
RECRUITMENT RATE (PEOPLE/MONTH)
LEAVING RATE (PEOPLE/MONTH)
STAFF IN TRAINING (PEOPLE)

Figure 5.19 SRL model with a totally exogenous leaving rate

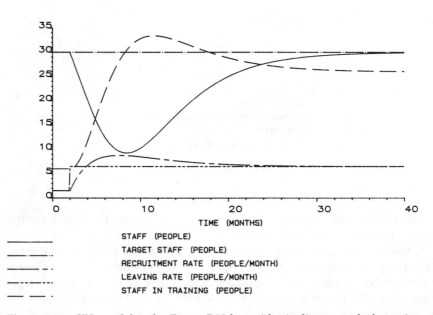

STAFF (PEOPLE)
TARGET STAFF (PEOPLE)
RECRUITMENT RATE (PEOPLE/MONTH)
LEAVING RATE (PEOPLE/MONTH)
STAFF IN TRAINING (PEOPLE)

Figure 5.20 SRL model as for Figure 5.19 but with pipeline control of recruitment rate

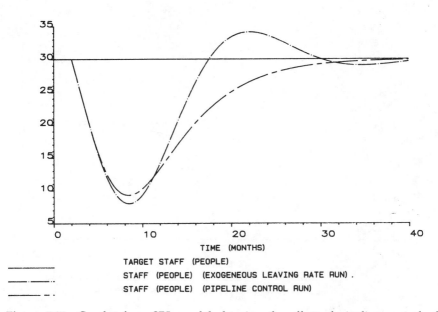

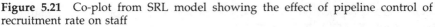

Figure 5.21 Co-plot from SRL model showing the effect of pipeline control of recruitment rate on staff

In fact the oscillations are not as severe as expected. One reason for this is the effect of the policy link between staff and leaving rate. When the leaving rate is suddenly increased the staff level falls but, this in turn causes the leaving rate to fall. It is of interest to experiment with this link and Run 10 in Table 2 in Appendix 1 amends the model to eliminate the link. The leaving rate now becomes purely exogeneous.

Figure 5.19 shows the results from Run 10 for the same variables as Figure 5.18. It will be seen that greater oscillations do occur.

The influence diagram of Figure 5.17 and the equations of Run 11 in Table 2, Appendix 1 also incorporate an experiment to try to eliminate some of the fluctuations. This policy experiment was set up to try to take into account the number of staff in training when defining the recruitment rate. A target STAFF_IN_TRAINING was defined, equal to the product of the average leaving rate and the training delay. The discrepancy between the target and actual number of staff in training was then divided by the recruitment delay and used as an additional component of the recruitment rate equation. Such a policy is generally referred to as proportional plus inertial plus pipeline control.

The results from Run 11 are shown in Figure 5.20. It will be seen that making the recruitment rate a function of the number of staff in training

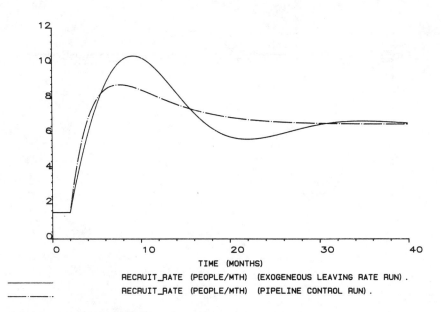

TIME (MONTHS)

—————————— RECRUIT_RATE (PEOPLE/MTH) (EXOGENEOUS LEAVING RATE RUN) .

—— · —— · RECRUIT_RATE (PEOPLE/MTH) (PIPELINE CONTROL RUN) .

Figure 5.22 Co-plot from SRL model showing the effect of pipeline control of recruitment rate on the recruitment rate

changes the behaviour of the system. The staff level returns to the target value with no overshoot and the recruitment rate stabilises at the new leaving rate with much less overshoot than in Figure 5.19.

Figures 5.21, 5.22 and 5.23 give co-plots of three model variables (staff, recruitment rate, leaving rate and staff in training) to highlight the improved behaviour. Pipeline control refers to the act of taking into account resources already committed (that is, those in the 'pipeline') when determining the rate at which resources are to be committed in subsequent time periods. In a production system the amount of resource in the pipeline would be referred to as work in progress.

SUMMARY

The purpose of this chapter has been to introduce a number of features of System Dynamics simulation and policy analysis. Policy analysis will be seen to be an iterative and self perpetuating procedure, where the behaviour of the system under one policy will lead to insight and to ideas for policy change.

The next three chapters will develop much more complex models.The emphasis in these models will again be on their use in developing systemic

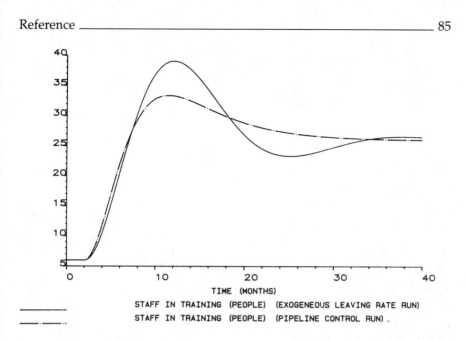

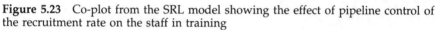

Figure 5.23 Co-plot from the SRL model showing the effect of pipeline control of the recruitment rate on the staff in training

understanding, solving problems and designing policies for improved system behaviour.

Selected points of equation formulation will be discussed based on the range of topics presented in this chapter.

REFERENCE

Dangerfield, B. and Vapenikova O. (1987) DYSMAP2 User Manual, University of Salford.

Chapter 6

A Case Study in Long Term Company Evolution

INTRODUCTION

This chapter presents an example of how Quantitative System Dynamics can be used to assess the effects of pre-defined strategies on the profits of a company.

The company used in the case study (CIR) was a medium sized subsidiary of a large group employing some 500 people and producing quality products almost exclusively for a government department. The company had an excellent image and had traditionally enjoyed a privileged position in the market with a certain security of contract and flexibility to regulate prices. Recent changes by its major customer had, however, resulted in a tightening of procurement procedures, resulting in a higher level of price consciousness and an encouragement of supply competition.

CIR had in recent years (along with its major traditional competitors) seen a decline in volume growth and allowable price increases with a resultant decline in operating profit from 22% to 16% over the last five years. This figure was forecast to fall to 12% over the next two years unless measures were taken to prevent it. The policy of the major customer was not anticipated to change in the foreseeable future, hence, borrowing, which might have carried the company through a short difficult period, was not seen as a viable option.

Company strategy to counter the situation in the recent past had been centred on two policies:

(1) Maintenance of orders by reinforcement of the company's image as a quality supplier with a good marketing and delivery record.

(2) Reduction of costs by investing in new, high technology machinery which would both increase workforce productivity (and hence reduce the number of employees) and save physical space.

These measures had been quite successful and CIR had maintained a higher percentage operating profit than its major competitors. However, absolute profitability had clearly reduced and there was concern about the overall future prospects of the company.

In order to evaluate the longer term seriousness of the current trend in profitability and to create a medium for analysis of alternative strategies, a model of some of the key variables of the company was developed. The structure and assumptions of this model will be described, followed by its application to generate alternative scenarios for the company's evolution.

INTERPRETATION OF THE PROBLEM IN TERMS OF FEEDBACK LOOPS

The situation described in CIR clearly has a reference mode of behaviour. This is one of declining profits arising from a decline in revenue as price and volume of business are constrained by the customer. These factors are external to the company and cannot be changed unless alternative markets and customers are found. In the situation described such changes are not possible. Consequently, there is no feedback loop underlying the decline in revenue which can be identified and manipulated by the company.

What the company can control is the other side for the profit equation, that is, to reduce costs by investment and hence compensate for the profit decline. This means establishing a positive feedback loop involving profit, investment and costs as shown, without defining variable types, in Figure 6.1.

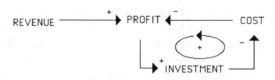

Figure 6.1 A feedback interpretation of the CIR situation

IDENTIFICATION OF RESOURCES AND STATES

The relevant resources, states and resource flows for the company, relating to the problem described, are given in Figure 6.2. It should be noted

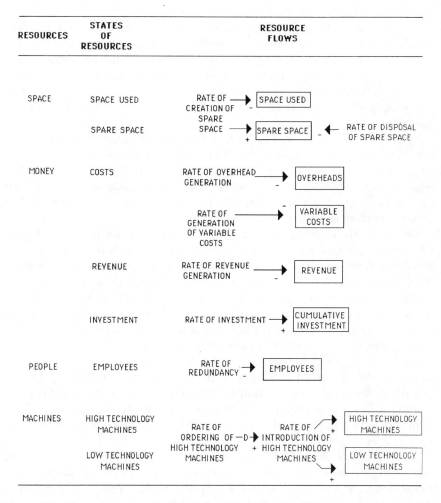

Figure 6.2 Resources, states and resource flows for the CIR case

that these were identified from a consideration of the processes at work within the company and the current and proposed strategies. The following thinking was involved. Four resources can be identified directly from the problem as described. These are space, money, people and machines.

Within the resource space, two states can be identified: space used and spare space. It is intended to convert the former to the latter by a strategy of investment in technology. However, space must still be paid for even if it is transferred from used space to spare space. Savings are only generated if this spare space is subsequently sold or rented, hence, a resource flow can be formulated involving space used, spare space, the rate of creation of spare space and the rate of disposal of spare space.

It is of interest to note that space could be considered in this problem simply as a cost. Ultimately, every resource can be reduced to monetary terms and, in general, care must be taken when formulating models not to reduce them to simply financial models of the spreadsheet type. One of the key characteristics of system dynamics is that of isolating individual resources and, at least initially, it is often best to identify all resources uniquely.

Within the resource money four states can be identified. These are the individual cost categories (overhead and variable costs), revenue and investment. Initial thinking centred on the fact that it would be useful to separate out overhead costs, variable costs and revenue as levels in their own right, particularly as it was likely that each might be subjected to separate percentage inflation factors.

This contrasts with the approach in Chapter 3 for the ESS case (Figure 3.2) where each cost and revenue rate was integrated directly into a profit variable defined as a level. Either approach is valid from a qualitative point of view. However, the choice of the approach used here is preferable from a quantitative point of view and this choice indicates the merits of thinking ahead to the equation formulation stage of model construction even during the diagram conceptualisation stage. The merits of the decision will become more apparent in the section on equation formulation.

The fourth state of money of interest in the problem is the cumulative money invested which, together with the rate of investment, creates a further resource flow. Actually, although cumulative investment is important in an accounting sense, it is the rate of investment which is of primary interest to the model structure. This is because money is actually used as it is invested.

The major state of interest within the resource of people is the number of current employees of the company. A resource flow can be constructed from this state and its rate of reduction, as implied by the strategy of investment in technology, which is aimed at reducing costs such as wages.

As with space, it is possible to consider two states of the resource machinery. These are existing (low technology machines) and new (high technology machines). This is because each will have a different productivity associated with them. The resource flow associated with machines must, obviously, involve a rate of introduction, which replaces low technology machines by high. Additionally, it is useful to realise that there is a significant time delay between ordering and introducing new machines.

CONSTRUCTION OF A COMPOSITE QUALITATIVE MODEL

Figure 6.3 presents a composite influence diagram of the model developed by combining the resource flows in Figure 6.2. This will now be explained. The resource flows associated with costs are considered first.

Further thought at this stage of model development led to both variable and overhead costs being split into two subcomponents—wages and other variable costs. Wages costs are separated out since they are based on the number of employees, which creates a direct link to the resource flow associated with employees. Overhead costs are also split into

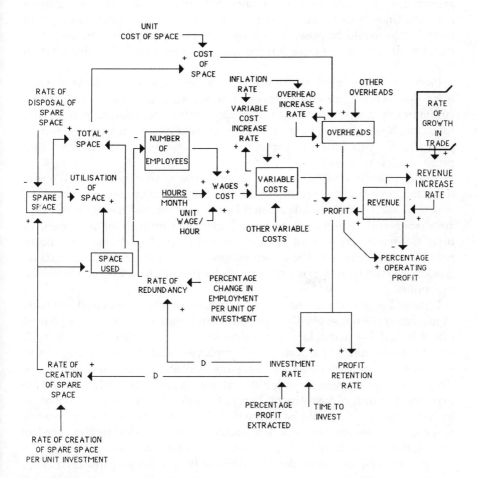

Figure 6.3 Influence diagram for CIR

two components: those relating to the total cost of floor space used by the company and other overheads. Again, this creates a link to the resource flow associated with the cost of space.

For further realism, it was agreed that inflation would be represented in the model and applied to both variable and overhead costs. In order to do this an increase rate was defined for both the variable and overhead cost levels as a function of the costs themselves and the percentage inflation rate. Note that the inflation rate is not a true rate as defined in System Dynamics. Percentage factors are commonly referred to as rates. Another example is the interest rate.

It will be seen from Figure 6.3 that revenue generation is defined as a function of an exogenous variable to the model. This is the anticipated growth rate in trade from the company's major customer. This growth rate combines both volume and price growth. It is assumed that a small growth rate would be possible as long as a strong marketing presence was maintained. The rate of growth in trade is modelled exactly like the inflation rate.

Profit is represented in Figure 6.3 as an auxiliary variable created from the levels representing revenue and costs. As such it will change cumulatively and continuously over any defined simulation period. The intention, however, was to sample it at each year end to measure the annual profit. Such a procedure corresponds to the company practice of extracting profit at each year end.

Further, profit is split into two components. One is retained profit, the other investment. Investment, is directed primarily into new machine procurement, hence, a link is created between profit and the resource flows associated with investment and machines. The number of machines, in turn, influences the rate of creation of spare space and also the rate of reduction of employees. These two factors influence overhead and variable costs, respectively, hence, the investment loop of Figure 6.1 is created within the model.

It should be noted that the states of cumulative investment and numbers of machines are not explicitly represented in Figure 6.3. Rather, investment rate is linked, via two delays (equivalent to the machinery ordering delays), directly to the rate of redundancy of employees and the rate of creation in floor space. This is because the company had already derived figures for the percentage changes in employment and spare space, which could be expected per unit of investment, hence, the intervening resources could be left out. The reader should be aware that these shortcuts were employed to make the best use of company data and to increase model credibility in the eyes of the company staff, at the expense of more rigorous modelling.

The time base of the model was defined in months and the percentage operating profit (the major variable of concern to the company) was calculated as the percentage ratio of the cumulated profit to the cumulated revenue at each year end.

EQUATION CONSTRUCTION

The model shown in Figure 6.3 was translated into a set of simulation equations using the DYSMAP2 programming language. Table 1 in Appendix 2 contains a full, documented listing of the model representing each sector of the influence diagram shown in Figure 6.3 (profit calculation, investment generation and its effects, overhead costs, variable costs and revenue generation).

It will be apparent that, unlike the Staff Recruitment and Leaving (SRL) model in Chapter 5, this model was constructed in a more traditional way by including all equations prior to the run statements and switching sectors of the model on and off using constants between run statements.

Although, this approach does not make full use of the DYSMAP2 facility to define new variables and equations between run statements, it reflects the way in which the model was created in practice, that is, all equations were constructed for all sectors and then decisions made as to which should be switched in and out during experimental runs.

In order to provide the reader with more detailed examples of equation formulation in System Dynamics, and to provide a basis for analysts to understand the model, a full explanation will be given of the creation of equations for one sector of the model. This is the variable cost sector which contains 18 active equations out of a total of 39 for the full model.

Those readers more interested in the use of the model, rather than the detailed formulation, should skim over this section.

Figure 6.4 presents a list of equations extracted from Appendix 2 for the sector of the model concerning variable costs.

It will be immediately apparent that there is somewhat more detail in the equations of Figure 6.4 than was apparent in Figure 6.3. The detail arises out of a need to represent the microstructure of the system. In the case here, this is mainly made up of more explicit definition of the components of wages costs and the variables necessary to calculate the annual variable cost. Each equation in Figure 6.4 will now be explained in turn.

Equation 6.1 calculates the wages cost per month (WC) as the product of the number of employees (E) and the wages cost per month per employee (MWR). The latter is an auxiliary variable defined in equation 6.2 as the product of the unit wage rate (UWR), the hours worked per day (HPD) and the days worked per month (DPM). These factors are constants as specified in equations 6.3, 6.4 and 6.5 respectively.

Equation 6.6 is the level variable representing variable costs (VC) in £. It

Equation	Equation Number
R WC.KL=MWR.K*E.K	6.1
A MWR.K=UWH*HPD*DPM	6.2
C UWH=2	6.3
C HPD=8	6.4
C DPM=30	6.5
L VC.K=VC.J+DT*(WC.JK+OVC.JK+VCIR.JK)	6.6
N VC=4326000	6.7
R VCIR.KL=INF*VC.K	6.8
R OVC.KL=115000	6.9
C INF=0.0034	6.10
A AVC.K=OVC1.K-OVC.K	6.11
A VCSF.K=VC.K-OVC1.K	6.12
L OVC1.K=OVC1.J+DT*PULSE(VCSF.J/DT,IPT,PIN)	6.13
N OVC1=VC	6.14
L OVC2.K=OVC2.J+DT*PULSE((0VC1.J-OVC2.J)/DT,IPT,PIN)	6.15
N OVC2=0	6.16
C IPT=12	6.17
C PIN=12	6.18

Documentation (M=months, P=pounds, D=days, H=hours, E=employees)

D	AVC=(P). . .	ANNUAL VARIABLE COSTS
D	DPM=(D/M).	DAYS PER MONTH
D	HPD=(H/D) .	HOURS PER DAY
D	IPT=(M) . . .	INITIAL PULSE TIME
D	PIN=(M). . .	PULSE INTERVAL
D	INF=(1/M) .	INFLATION RATE
D	MWR=(P/M/E)	MONTHLY WAGE RATE
D	OVC=(P/M) .	OTHER VARIABLE COST
D	OVC1=(P) . .	OLD VALUE OF VARIABLE COSTS 1
D	OVC2=(P) . .	OLD VALUE OF VARIABLE COSTS 2
D	UWH=(P/H/E)	UNIT WAGE RATE
D	VC=(P) . . .	VARIABLE COST
D	VCSF=(P) . .	VARIABLE COSTS SO FAR IN YEAR
D	VCIR=(P/M).	VARIABLE COST INCREASE RATE
D	WC=(P/M) .	WAGES COST

Figure 6.4 DYSMAP equations for the variable cost sector of the CIR company model

has an initial value given in equation 6.7 and is increased by the wages cost rate (WC), other variable costs (OVC) and the variable cost increase rate (VCCR), all having units of £/month. The variable cost increase rate changes the variable costs by an inflation factor (INF), which is defined as a proportion of variable costs (see equation 6.8). The inflation factor is given in equation 6.10 as a constant of 0.0034 per month (or approximately

4% per year). Other variable costs (OVC) are defined at a constant rate by equation 6.9.

The variable cost, by definition of being a level, represents the *cumulative* variable cost over the whole simulation run. What is required is to calculate the annual variable cost in each year. This can then be taken along with the annual overhead cost and revenue to calculate the annual profit and the annual percentage operating profit for the company. The annual profit must also be known in order to calculate the investment rate.

The method used here to calculate the annual variable cost is one often employed in System Dynamics models (Coyle 1977) and further details can be found in this source.

The situation is described graphically in Figure 6.5. What is required is that the variable cost level be sampled at the end of each of the two most recent complete twelve month periods in the model. The difference between these samples gives the annual variable cost (AVC), defined in equation 6.11.

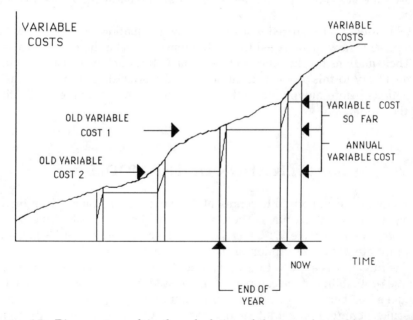

Figure 6.5 Diagram to explain the calculation of the annual variable cost in the CIR model

The SAMPLE function in DYSMAP2 is convenient to take one sample of a time series. However, this does not work if two samples are required. In order to take two samples, two 'old values', of the time series must be calculated. These are the old variable cost 1 (OVC1) and the old variable cost 2 (OVC2) shown in Figure 6.5. It should be noted that although OVC1 and

OVC2 are values of VC at past time points, they are carried forward in the software and are current values at the time point labelled NOW. These old values are calculated by using the PULSE function to make rapid changes to them during the simulation interval (DT) immediately following the end of each 12 month period.

The simulation interval is a function of the simulation software and not the model and should never, in normal circumstances, occur on the right-hand side of equations other than level equations. Its use here is an exception to that rule. Whenever such exceptions to modelling practice are used, it is vital that the reasons are clearly understood by the modeller.

Old value 1 is calculated as a level variable which is increased by a pulse of the 'variable cost so far' (VCSF) divided by DT at each year end. The variable cost so far is defined in equation 6.12 and is in fact the difference between the actual variable cost and OVC1. The equation for OVC1 is given in equation 6.13, the initial condition for it is in equation 6.14 and the initial pulse time (IPT) and pulse interval (PIN) are given in equations 6.17 and 6.18.

Old value 2 is calculated in a similar way in equations 6.15 and 6.16, but the pulse is now made equal to the difference between the two old values.

The equations for the other sectors of the CIR model are constructed in a similar way to this sector and the interested reader should examine those equations before moving on to the next section concerning the application of the model.

APPLICATION OF THE MODEL

The model was initialised to represent the current situation as given by the company's annual accounts. These figures for the base year of the model are presented on page 97. Other parameter values used will be given as appropriate in the description of the experiments on the model.

The model was applied to demonstrate the effects of a number of factors on the profitability of the company. It is stressed that the experiments on the model given here are not exhaustive and represent only a small selection of possible permutations of alternative assumptions and company strategies which could be tried out. Although the results of the experiments were useful in their own right, the purpose of the study was wider than these immediate results. The aim was firstly, to create a medium by which to focus attention on potential trends in company performance. Secondly, to develop communication and to encourage debate and, thirdly, to demonstrate the use of the model as a test bed for managerial thinking.

	Annual (£000)	Monthly £000)	% of sales
Sales	15763	1314	100.0
Overheads Costs:			
Space	2000	167	—
Other	6900	575	—
Total	8905	742	—
Variable Costs:			
Wages	2940	245	—
Other	1380	115	—
Total	4326	360	—
Operating Profit	2532	212	16.06

Run 1, Base Case (Inflation = Growth in Revenue)

The model was first run in equilibrium with the annual rate of inflation equal to the annual rate of growth of revenue (set at 4% per annum) and without any reinvestment of profit. This was achieved by setting the investment switch (ISW) to zero in the generation of investment sector (see model listing in Appendix 2). The results of this base run are given in Figure 6.6. This shows the evolution of the annual percentage operating profit, percentage of space utilised and number of employees over a 12.5 year period (150 months).

The purpose of this run was, essentially, a model validation exercise, both to test for false dynamics and model consistency. As expected, equal growth in both costs and revenue results in all three variables remaining constant as demonstrated by the horizontal lines in Figure 6.6.

Run 2 (Inflation Exceeds Growth in Revenue)

The model was next run under the more realistic assumption that the rate of inflation would, in the foreseeable future, exceed the rate of growth of revenue. Inflation was again fixed at 4% per annum and the following annual rates of growth set for revenue:

Years (beyond base year)	1	2	3	4	5	6	7	8	9	10	11	12
Revenue Growth Rate (%)	2.8	2.8	2.6	2.6	2.6	2.6	2.5	2.5	2.4	2.4	2.3	2.3

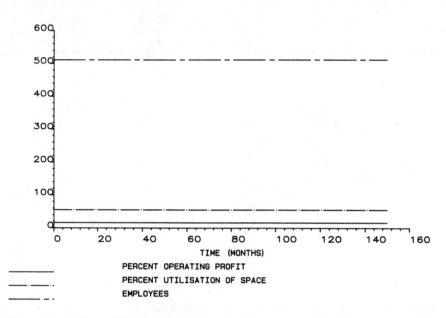

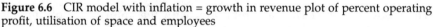

Figure 6.6 CIR model with inflation = growth in revenue plot of percent operating profit, utilisation of space and employees

Again no reinvestment was considered. Figure 6.7a shows a plot of the outcome of this effect over time for the same variables as Figure 6.6. The stepwise nature of the curve for annual percentage operating profit arises from the discontinuities introduced in the sampling of costs and revenue.

As can be seen in Figure 6.7a a serious decline in annual percentage operating profit clearly takes place, with this variable becoming negative by the end of the simulated period. From a validation point of view, this run demonstrates the ability of the model to create the mode of behaviour being currently experienced, and anticipated for the future, by the company.

Apart from their role in validation, however, the results of runs 1 and 2 taken together, actually suggest the root *cause* of the declining profitability trend in CIR. This can be simply stated as follows. If future revenue growth is likely to be constrained by the major customer of the company to less than 3% per annum, then this is unlikely to cover the increase in operating costs due to inflation. A comparison of annual revenue between runs 1 and 2 is shown in Figure 6.7b.

From a modelling point of view the outcome of runs 1 and 2 are of great interest. The model was originally constructed around some of the company's intended strategies. This was perhaps a mistake, in that the model incorporated the management solutions of what *might* be, before

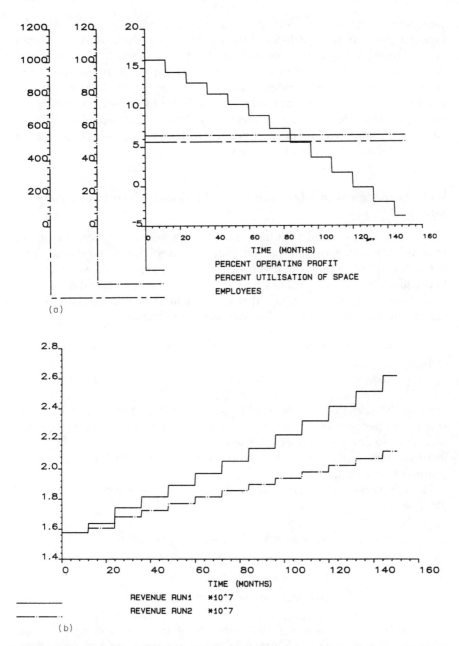

Figure 6.7 (a) CIR model with inflation exceeding growth in revenue. Plot of percent operating profit, utilisation of space, and employees, (b) CIR model: comparison of revenue between run 1 (inflation=growth in revenue) and run 2 (inflation exceeds growth in revenue)

considering what *was*. In fact, inflation was a factor only included in the model to make it more realistic, at the quantification stage of development.

As shown in run 2, the situation without investment is clearly one of degenerating profits arising from the company's major supplier restricting volume and price growth rates to less than the anticipated growth rate in inflation. This is the inherent mode of behaviour of the company under the assumptions made. It is to be expected, therefore, that implementation of any of the company's planned strategies will only slow this degenerative mode of behaviour, and not reverse it.

Run 3 (Investment in High Technology Machines (Current Company Strategy))

The model was next run with the same revenue/inflation differential given in Run 2, but incorporating the company's strategy for reinvestment. This was achieved by setting the investment switch (ISW) to 1. It was assumed that half of the annual profits could be reinvested each year and that this would have two effects. Firstly, that investment in new technology would reduce the workforce at the following rates over time:

YEAR (beyond base year)	1	2	3	4	5	6	7	8	9	10	11	12
Reduction in employees per £ million invested	30	30	25	25	20	20	20	20	10	10	5	5

Secondly, that investment would reduce the floor space used by 0.006 square feet per pound invested per year. However, it would simply transfer this space to spare floor space, for which similar overheads cost would be incurred. Hence, the space utilisation would decline but the total cost of space would be unchanged.

The following initial values were extracted from the company accounts for this run.

Number of Employees	511
Average Wage Rate	£2/hour
Hours Worker Per Day	8
Days Worked Per Month	30

Figure 6.8a shows a plot of the outcome of this run over time for the same group of variables as Figure 6.7a and Figure 6.9b shows a comparison of variable costs between this run and run 1.

It will be seen from Figure 6.8a that the number of employees is reduced from 511 to 263 over the full period simulated and that the percentage operating profit is maintained at a higher value than in Run 2. This attains a value of 5.1% after 12 years. The effect is also, marginally, to reduce space utilisation to 51.5% over the whole period. In other words, as expected, a reinvestment strategy of the magnitude used here will offset the imbalance between revenue and inflation, but only to a limited extent.

Moreover, the effect of the investment becomes less and less through time as reflected in the graph for annual percentage operating profit in Figure 6.8a. The rate of decline of this variable increases over time. This is not only because of the inbuilt reduction in the marginal return of investment on employees, but *also* due to the feedback between profitability and investment built into the model. Declining profitability in each year means less investment is available in each year.

The annual profit generated after 12 years is £1.01 million compared with £2.53 million in the base year. Annual investment figures for these two points in time are £0.5 million and £1.25 million respectively.

An alternative approach to modelling this particular company problem might have been to create a spreadsheet model. The usual approach in such models is to create independent inputs to the model in successive time periods. This can be considered as formulating them in an 'open loop' rather than 'closed loop' or feedback mode. Typically, such a formulation would not have linked investment to profit *and* profit to investment. The result would have been that the predicted percentage return on investment would have appeared more healthy. This point emphasises the importance of making models close looped and taking feedback into account.

It can be concluded that, in a position of serious revenue–inflation imbalance, it is important to reduce costs by increased productivity, but ultimately such measures cannot reverse the underlying downward trend in profitability resulting from the imbalance between revenue and inflation.

Run 4 (Reduction of Unused Space)

Given the previous conclusion some more radical approaches by which to improve profitability are obviously necessary. Apart from seeking new, more flexible markets the only alternatives centre around more efficient use of existing resources, which focuses attention on the issue of spare floor space. The company currently owns 2000 square feet of floor space and account this at £1 per square foot. The current percentage space utilisation is 55%.

The purpose of Run 4 of the model was, therefore, to quantify the percentage operating profit which would result from achieving a reduction in spare space. The following monthly reductions in spare space were used. These were based on estimates of the rate at which space might be transferred, as it became available, to other agencies within the company's

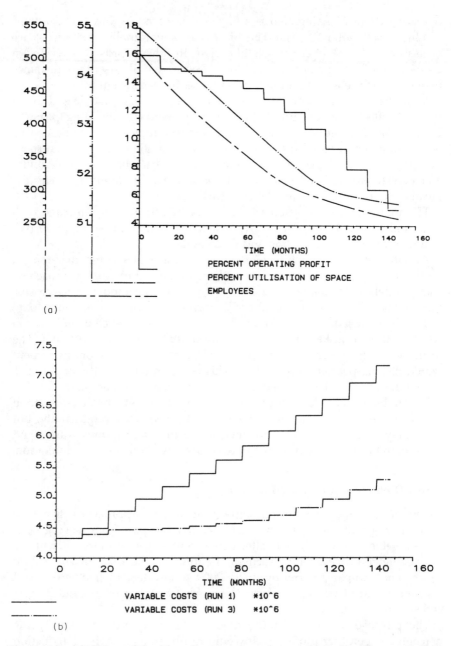

Figure 6.8 (a) CIR model with reduction of employees. Plot of percent operating profits, percent utilisation of space and employees (b) CIR model: Comparison of variable costs between run 1 (inflation=growth in revenue) and run 3 (reduction in employees)

group. Such transfer would simply eliminate the overhead cost associated with the space. No account was taken of additional revenue or rent arising from the transfer.

YEARS (beyond base year)	1	2	3	4	5	6	7	8	9	10	11	12
Saving in unused space per month (square feet/month)		700	700	700	700	500	500	500	500	300	300	300	300

The reductions were applied each month in the model to decrease the total cost of spare space by changing the spare space switch (SSW) from zero to one in the 'investment in space saving' sector of the model (see Appendix 2).

Figure 6.9a shows the outcome in terms of the previously plotted variables. It will be seen from Figure 6.9a that the effect of this strategy on the percentage operating profit is dramatic. This figure falls but only to 11.5% per annum at the end of the simulated period. Utilisation of floor space in this run rises to 85% suggesting that there is still scope to improve on these profits.

The high magnitude of the effect of saving floor space can be explained in terms of its cumulative effect on overhead costs. A removal of 700 square feet of floor space in month 1 results in overhead savings in month 1 of £700. A removal of another 700 square feet in month 2 results in an overhead saving in month 2 of £700 (from the month 2 reduction) *plus* £700 from the month 1 reduction. In month 3 the saving is £2100. The cumulative effect of this process gives a saving in annual overheads of over £1 million at the end of 12 years. A comparison of overhead costs between runs 1 and 4 is given in Figure 6.9b. The additional profit arising out of the reduction in overheads in turn facilitates a greater investment rate each year and, hence a further reduction in employees and, hence, in variable costs. A comparison of variable costs between runs 3 and 4 is given in Figure 6.9b. The feedback effect here is again important.

It was, therefore, concluded that the effect of savings on unused space was certainly a worthwhile issue to pursue. Surprisingly, the magnitude of this saving was sufficient to more than counter the decline in operating profit in the medium term. However, the downward trend was re-established in the longer term.

Figure 6.10 shows a comparative plot of the figures for the percentage return on investment, from each of the four experiments described and Figure 6.11 shows a modified version of the underlying feedback structure of the model shown in Figure 6.1. This is drawn after the analysis and emphasises the way in which the exogenous inflation rate and growth in customer sales–price compete to influence profit. It also demonstrates how

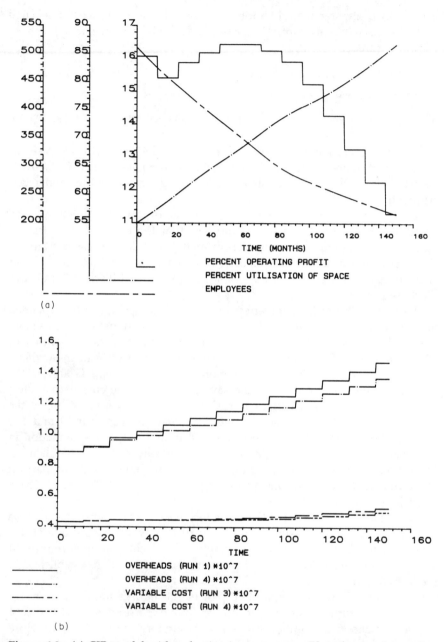

(a)

(b)

Figure 6.9 (a) CIR model with reduction in spare space. Plot of percent operating profit, percent utilisation of space and employees (b) CIR model: Comparison of annual overhead costs between runs 1 and 4 and annual variable costs between runs 3 and 4

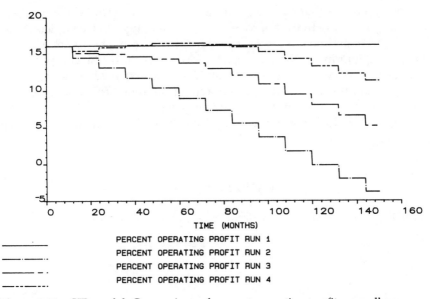

Figure 6.10   CIR model: Comparison of percent operating profit over all runs

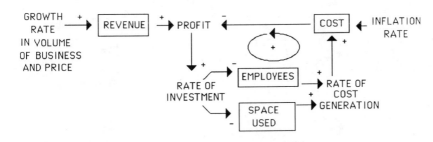

Figure 6.11   Retrospective feedback interpretation of the CIR situation

profit can be fed back to reduce costs, but that this is only a compensatory measure to counter any imbalance between revenue and inflation growth.

From a modelling point of view it should be pointed out that two of the influence links in Figure 6.11 only work in one direction of change. Increases in investment in technology will reduce the number of employees and the space used but not vice versa.

CONCLUSIONS

This study involved the creation of a model in mainly financial terms where strategies had been defined, but needed quantitative testing. The model was conceptualised largely on the basis of these strategies, which involved investing in productivity improvements. It was only after completing the model and experimenting with it in the absence of reinvestment and in a classical equilibrium situation, that the real problem of the company was identified. This was that the company's anticipated revenue growth rate was less than the anticipated growth rate in inflation. This understanding enables the cost reduction strategies to be viewed in a new light: they could, at best, only reduce the effect of, and never totally cure, the underlying problem.

REFERENCE

Coyle, R.G. (1977) *Management System Dynamics*, Wiley, Chichester.

Chapter 7

A Case Study in the Control of a
Coal Transportation System

INTRODUCTION

With the advent of new technology using microcomputers for the centralised *retrieval* of information, the scope for totally automatic, real time control of large systems has been advanced. We are entering a period which, retrospectively, may well be seen as the era of control: where, because of the cost and limit of resources, it is both necessary and feasible to design systems in terms of the way in which they will be operated, as well as in the traditional terms of their capacity requirements. Before such comprehensive design can be undertaken, however, compatible advances are required in information *usage* technology, that is, in the technology to design operational control rules for systems.

This chapter deals with an example of how a System Dynamics model can be used to quantitatively design control in a system and to quantify the savings in the physical capacity of the system made possible by the improved control.

If the capacity of a system is infinitely large (that is, the upper limit of the quantities which can pass through it are not limited), control is totally unnecessary. This is because process flows will never be inhibited. However, such systems do not exist in practice and, in real systems where capacity is limited, control becomes more and more important and can be used as a substitute for capacity.

The problem described is taken from the coal mining industry and relates to the design of large scale underground conveyor belt systems used for the clearance of coal from mines.

The model used in this chapter was developed using the STELLA software.

BACKGROUND TO THE PROBLEM

The majority of coal mines utilise conveyor belt systems to transport coal from the working coal faces to the surface. The major problem in designing the capacity of such systems is that coal face output rates fluctuate over time, due to variations in the shift patterns in operation, variations in the coal cutting speed (due to geological changes) and the reliability of coal face machinery. Further, since reserves of coal are depleted by extraction, the layout of coalfaces in a mine is a geographically dynamic phenomenon.

The capacity of coal clearance systems must be designed to cater both for short term fluctuations and longer term changes in coal production rates. As a result of major developments in coal mining technology over recent years and the consequent trend towards concentrating production on fewer and larger coalfaces the situation where colliery coal clearance systems are working at or above their design capacity is frequently encountered. Consequently, attention is now being focussed on the use of more sophisticated control of these systems in order to make the best use of existing capacity. The feasibility of such control has been enhanced by technological advances in the development of minicomputers and microcomputers for real time control.

The installation of computers is currently taking place on an increasing scale at colliery level, and these are being used to monitor and display up to date information on both coal clearance and many other underground systems in central control stations. The implementation of action based on this information is, at present, largely manual, but the ultimate potential of these applications is that decision rules or control policies can be automated, allowing control actions to be fed directly back to the operations.

The fundamental difficulties in attaining this potential in any information and control system are those of determining which information sources to monitor, and what form of control rule to use. This presents somewhat of a dilemma, because control rules cannot be formulated unless a choice of information has been made and it is difficult to choose which information to monitor unless the benefits of using it in control rules have been assessed.

In the past much effort has been put into improving the operation of coal clearance systems by using discrete entity simulation. By definition, such studies have been concentrated on capacity planning because these techniques do not easily facilitate control design.

THE PROBLEM

The specific problem posed was that of designing the capacity (bunker and conveyor belt sizes) and control rules for a new three bunker coal clearance system in a large drift mine. Each bunker was to be fed by a single coalface and would discharge on to a conveyor belt, which would transport the coal up the drift to the surface of the mine.

As the system was new, there was no specific reference mode for its actual behaviour. However, there was an overall objective to achieve. This was to move as much coal from the coalfaces to the surface as possible each day and to avoid any necessity to cease production at the coalfaces because of congestion in the clearance system.

The physical flows of the system studied are shown in Figure 7.1. The overall aim was to develop and test alternative policies for the rates at which the bunkers should discharge coal onto the conveyor belt, measured against the efficiency criterion of the ratio of cumulative coal output cleared to the surface, over a given period of time, to that potentially available from the three coalfaces.

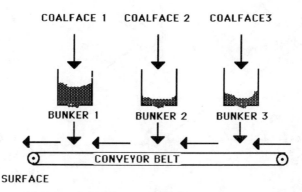

Figure 7.1   General layout of situation modelled

THE CONCEPTUALISATION OF THE MODEL

This system is relatively easy to conceptualise as a System Dynamics model. Figure 7.2 shows a composite pipe diagram of the model, of the type introduced in Chapter 2.

There is a single resource to consider. This is coal and the main states of interest within this are the quantities of coal in the bunkers. The bunkers, by definition, are stocks and can be represented by levels. Further, the coalface output rates clearly link to the bunker input rates, which in turn must feed the bunkers and the bunker discharge rates must clearly deplete the bunkers. Hence, three structurally identical resource flows are created, one for each bunker. The bunker input rates must be separated from the coalface

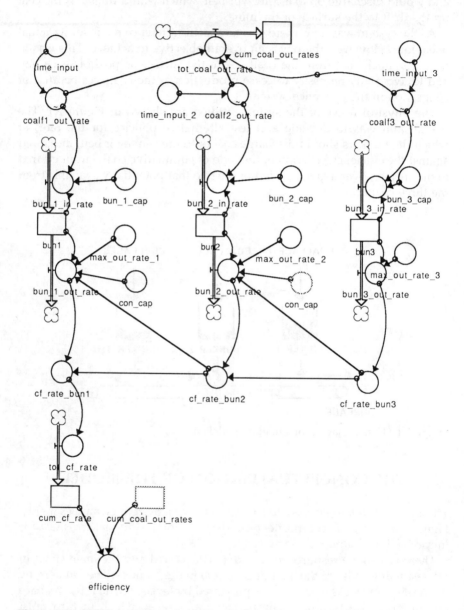

Figure 7.2 STELLA diagram for Policy I of coal clearance model

output rates, since the latter may, of course, be less than the coalface output rates, if the bunkers are full.

The structure of the bunker discharge rates in Figure 7.2 correspond to those used in the first set of policy experiments and are described in detail in a later section.

From a model construction point of view, it is important to note that the coal on the conveyor belt is not explicitly modelled as a state of the coal resource. The bunkers are assumed to discharge onto the conveyor belt very close to one another, in a preset order (3,2,1). The coal flow rate on the conveyor belt at bunker 3 is modelled as as auxiliary variable, equal to the coal discharge rate from bunker 3. The coal on the conveyor belt at bunkers 2 and 1 (cf_rate_bun2 and cf_rate_bun1) are also modelled as auxiliary variables. These variables are the sums of the respective bunker discharge rates plus any coal flow already on the conveyor belt.

The coal flow rate at bunker 1, hence, represents the total coal flow rate on the conveyor belt (tot_cf_rate). The cumulative version of this is the cumulative coal flow rate (cum_cf_rate) at the surface. A similar summation of the coalface output rates gives the total coalface output rate (tot_coal_out_rate) and the cumulative version of this is the cumulative coalface output rate (cum_coal_out_rates) shown at the top of Figure 7.2. The ratio of this to the cumulative coal flow rate measures the efficiency of the system.

It is of interest to note that the situations encountered so far in this book have tended to view the performance of model in terms of the behaviour of one variable over time. It is often the case that clear measures of performance, such as the operating profit used by the CIR company in Chapter 6, do not exist and performance measures must be designed. In fact, System Dynamics models can often help with the process of designing suitable measures, in this case however, a ready made performance measure already exists.

It should also be noted that a good performance measure must summarise the behaviour of a model over a complete simulation run, that is, it should be cumulative over time. Further, performance measures are, as here, often compound variables, such as ratios of the two cumulated states.

THE STELLA SOFTWARE

A number of points of explanation are necessary concerning the use of the STELLA software in Figure 7.2. In STELLA all variables other than levels are depicted by the same type of circular icon, known as a converter. The STELLA software exerts a discipline on the modeller by only allowing information links on the computer screen to be made between certain variables. For example, it would not be possible for levels to be connected to other levels. However, STELLA does allow the direct connection of converters to converters and the modeller must be aware that this implies instantaneous information transfer. Such links exist in Figure 7.2. For example, between the coal flow rates at bunkers 2 and 3, and the discharge rates of bunker 1 and 2.

To simplify linkages STELLA also allows variables, once defined, to be replicated and 'ghosted' to any location on the computer screen. These ghosts appear with a dotted outline as seen in Figure 7.2 for the variables representing the cumulative coalface output rate (cum_coal_out_rates) and the conveyor belt capacity (con_cap).

EQUATION FORMULATION

Table 1 in Appendix 3 lists the STELLA equations created from Figure 7.2. Apart from the policy equations (shown in bold type which will be explained in later sections), these equations should be understandable to readers who have absorbed Chapters 4 and 5. Such readers are encouraged to examine the equations in conjunction with Figure 7.2. The first five equations are level equations (which are generated by the software directly from the diagram) and their associated initial conditions. The remainder are converter equations. Both types of equation should be recognised as being similar in construction to the DYSMAP2 equations, but without time subscripts. The main differences are the use of the more conventional computer science IF...THEN statements instead of CLIP functions, and the use of coordinates to represent the TABLE functions.

It should be noted that all the policy equations of the model in Tables 1, 2 and 3, Appendix 3 (which define the bunker discharge rates), are constrained. This is to ensure that the quantities in the levels they deplete (that is, the quantities of coal in the bunkers), never become negative.

There is always a danger in System Dynamics models that rates which deplete levels might take more from the levels than they contain at any time. This situation did not occur in the CIR model of Chapter 6 because the rate of reduction of employees and spare space never approached zero. However, it needs to be safeguarded against in the coal clearance model.

In general, the placing of constraints on rates which deplete levels can be achieved in two ways. The first is to allow a desired amount to be taken from a level right up to the point at which the quantity in the level reaches zero. An alternative approach is to define the depleting rate as a table function which progressively lowers the amount which can be taken from a level, as the quantity in the level approaches zero. The philosophy of this approach is one of introducing rationing of the resource in the level as it approaches zero. The correct modelling approach, in practice, depends on the particular circumstances encountered.

The first of the above methods is used in the coal clearance model. The following equation is taken from Table 1 of Appendix 1 to illustrate the approach.

```
bun_1_out_rate=IF(con_cap-cf_rate_bun2)>=max_out_rate_1
THEN MIN(max_out_rate_1,Max(0,bun1/DT))Else 0          (6.1)
```

where

`con_cap`	= conveyor belt capacity,
`cf_rate_bun2`	= coal flow rate leaving bunker 2 discharge point,
`max_out_rate_1`	= maximum discharge rate of bunker 1 and
`bun`	= quantity of coal in bunker.

This equation states that if the difference between the conveyor belt capacity and the coal flow rate already on the conveyor belt (as a result of bunker 2 discharge rate) is greater than or equal to zero (that is, if there is room on the conveyor belt), then it is desired to discharge bunker 1 at its maximum rate.

However, it may not be possible to achieve this maximum rate, if there is insufficient coal in bunker 1. Consequently, the actual bunker 1 discharge rate should be the minimum of this desired rate and the rate which can actually be taken out of the bunker in the next simulation interval, without the bunker giving negative, that is, `bun1/DT`. In the final form of equation 6.1, the situation is further safeguarded by replacing `bun1/DT` by the maximum of this or zero. This use of `DT` on the right-hand side of an equation is similar to the example presented in Chapter 6 and subject to a similar caveat.

EFFECTS OF ALTERNATIVE BUNKER DISCHARGE POLICIES

Design of Experiments

In order to produce a realistic pattern of the output of each coalface over time by which to drive the model, the coalface sectors of the model were designed to take into account shift working times and variations in the rate of coal cutting and stoppages due to machinery breakdowns.

This was achieved by incorporating real data in the form of graphs of the coal output rate from each coalface over time. The STELLA software facilitated the choice of this method since it allows such patterns to be drawn on the computer screen and automatically converts the graphs into table functions within the model. An alternative, to avoid the tedium of formulating lengthy time series inputs in other software, would have been to develop coalface programs separately using a series of step, clip and random functions. An example, of this latter type of formulation for coalfaces can be found elsewhere (Wolstenholme and Coyle 1980).

An example of a time series input used in the model for one coalface output rate (coalf1_out_rate) is presented in Figure 7.3. As shown in Figure 7.2, the actual coalface output rates produced from the time series inputs, were each fed directly into their respective bunkers (bun_in_rate) as long as the quantity in these remained below their capacities (bun_cap).

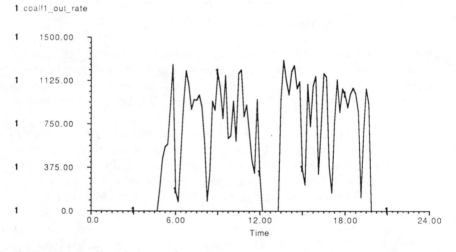

Figure 7.3 Coalface 1 output dynamics

All experiments and results presented here are based on varying the capacities of the bunkers, conveyor belt, the maximum bunker discharge rates and the bunker discharge policies. Common features of all the model runs were that the conveyor belt could operate throughout the working day, but that the coalfaces were operated over two six hour shifts. Each coalface was designed to produce 1000 tons per hour on average. Also the model was run in each case for one complete day (24 hours), starting from an equilibrium situation.

Each experimental run of the model was defined in terms of the parameters of the coal clearance system. These runs are listed on the left hand side of Table 7.1, against which various bunker discharge policies

Table 7.1 Overall system efficiencies for bunker discharge Policies I, II and III.

Definition of experiment		Overall system efficiency					
		Total belt capacity (tons/ hour	Maximum discharge rate of each bunker (tons/hour)	Capacity of each bunker (tons)	Bunker discharge Policy I (Fixed discharge)	Bunker discharge Policy II (Variable discharge)	Bunker discharge Policy III (Belt capacity allocation)
Block 1	Run 1	2000	1000	500	72.84	73.75	75.68
storage	Run 2	2000	700	500	55.73	68.20	75.04
bunker	Run 3	2000	1000	1000	76.30	77.86	83.00
system	Run 4	2000	1000	1200	78.47	80.73	85.60
Block 2	Run 5	2500	1000	150	64.59	64.79	69.54
surge	Run 6	2500	1000	500	73.17	84.48	90.78
bunker	Run 7	3500	1000	150	72.63	65.48	72.63
system	Run 8	3500	1000	500	98.46	86.16	98.46
	Run 9	3500	1000	750	100.00	89.91	100.00

were subsequently tested. The parameters defined in Table 7.1 can be seen to fall into two blocks.

In the first block of experiments, the belt capacity (2000 tons/hour) was chosen to be less than the sum of the base output rates of the three coalfaces. These parameters represent a storage bunker situation. Here, the purpose of the bunkers is to act as storage over the day to match coalface outputs with inadequate belt capacity. Four experiments were defined, which examined the effects of lowering the maximum bunker discharge rate from the base value of 1000 tons/hour to 700 tons/hour, and increasing the capacity of all bunkers from their base value of 500 tons to 1000 tons or to 1200 tons.

The second block of experiments represented a surge bunker situation, where the conveyor belt capacity was first set to 2500 tons/hour. At this level of capacity the conveyor belt is almost capable of dealing with the sum of the base output rates of the three coalfaces, but not the sum of their maximum rates. In this case, the purpose of the bunkers is to smooth out instantaneous fluctuations in coalface output rate over and above conveyor belt capacity. Obviously, less bunker capacity is required in these circumstances than in the storage bunker situation and experiments are hence defined in terms of 150, 500 or 750 ton bunker capacity. The conveyor belt capacity was subsequently increased to 3500 tons/hour. At this level of capacity the conveyor belt is capable of dealing with instantaneous peaks in coalface outputs, and the effects of various bunker capacities were examined.

Some rather obvious and crude bunker discharge policies were then tested using the model.

Policy I: Fixed Bunker Discharge Policy

It was assumed here that each bunker could only be discharged at zero or at its maximum discharge rate and that the latter would be used as long as there was coal available in the bunker and room available on the conveyor belt. This policy is specified in the diagram of Figure 7.2 and the equations highlighted in bold type in Table 1, Appendix 3.

Policy II: Variable Bunker Discharge Policy

It was assumed here that each bunker discharge rate could be set at any point between zero and the maximum discharge rate. Further, that it would be set in the same proportion as the bunker level to the bunker capacity. Again, the discharge rate was subject to coal being available in the bunker and room being available on the conveyor belt. This policy is specified in the modified diagram of Figure 7.4 and in the equations highlighted in bold type in Table 2 of Appendix 3.

The result of applying these simple bunker discharge policies to the runs previously defined are shown in Table 7.1.

Results from Policy I

With a conveyor belt capacity of only 2000 tons/hour (block 1 results) and a maximum discharge rate from each bunker of 1000 tons/hour, Policy I exhibits efficiencies between 72.84 and 78.47 (runs 1, 3 and 4). This is because, at best, only two bunkers can discharge at any one time. The difference between these runs demonstrates the increased efficiency resulting from increasing the bunker capacity and run 2 indicates the reduced efficiency resulting from a lowering of the maximum bunker discharge rate.

When the conveyor belt capacity is increased to 2500 tons/hour (runs 5 and 6 in block 2) no improvements in efficiency results because it is still possible to only discharge, at most, two bunkers at any time. Consequently, run 6 gives a similar result to run 1, and the efficiency for run 5 is reduced to 64.59 due to less bunker capacity being available.

Increasing the conveyor belt capacity to 3500 tons/hour results in improved efficiencies (runs 7 and 8) compared to the equivalent runs for a conveyor belt capacity of 2500 tons/hour, since all three bunkers can now be discharged together. Maximum efficiency at a conveyor belt capacity of 3500 tons/hour is achieved using bunker capacities of 750 tons (run 9).

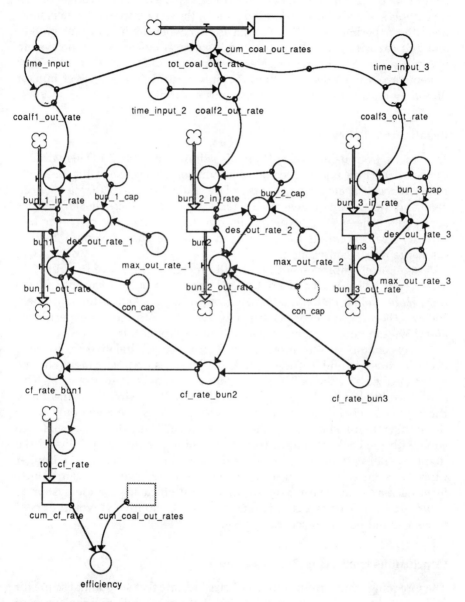

Figure 7.4 STELLA diagram for Policy II of coal clearance model

Under policy I the discharge rates from the bunkers are either zero or the maximum discharge rate, hence, there are times when some discharge could take place but does not and it is to be expected that the conveyor belt capacity is somewhat under-utilised. One of the most important conclusions from the experiments in this section is that when both sufficient bunker and belt capacity exist, the maximum theoretical efficiency for the system is attained (run 9). This result gives confirmation of the fact that if the total system capacity is adequate, then the only control necessary over bunker discharge rates is to deploy their maximum setting.

Results from Policy II

The results from using Policy II are also shown in Table 7.1. This variable bunker discharge policy overcomes the major deficiency of the previous policy, since discharge can now take place at any intermediate rate between zero and maximum. However, there is little change in the efficiencies achieved in the block 1 experiments due to the very low conveyor belt capacity. An exception to this is run 2 where the maximum bunker discharge rate is constrained.

A significant improvement in overall efficiency is achieved in run 6 at a conveyor belt capacity of 2500 tons/hour and a bunker capacity of 500 tons. Here, unlike Policy I, advantage can now be taken of the additional 500 tons/hour of conveyor belt capacity available over block 1 experiments and a 500 ton bunker capacity is sufficient to exploit this.

The exceptions to the improvements in efficiency, relative to Policy I, occur in runs 7, 8 and 9. These reduced efficiencies do in fact highlight one major weakness of the policy used. Once sufficient conveyor belt capacity exists, the bunker discharge rates under Policy II are simply a function of the bunker level and the maximum discharge rate is employed only when the bunker is full. Further, if the bunker capacity is almost adequate to cope with the coalface output (run 7), the bunkers are rarely full and the discharge rates attained are less than the maximum. Indeed, as the bunker capacity is increased, the discharge rate associated with a given bunker level decreases and the maximum theoretical efficiency for the system is never attained. This is a disadvantage of what, intuitively, would appear to be a sound bunker discharge policy.

Conclusions from using Policies I and II

The foregoing results from simple policies indicate that substantial scope for improvement in policy design exists and that a combination of the merits of Policies I and II should be the first step, that is, a policy is required which employs the maximum bunker discharge rate wherever possible and allows for intermediate discharge rates depending on the bunker levels.

Further, it is clear that both of the simple policies used represent priority policies. Examination of the distribution of coal losses by individual coalfaces highlights the problem. Coal loss refers to the coal lost when production is stopped at the coalfaces due to a bunker being full and no room being available on the conveyor belt. Results from Policies I and II indicate that the last coalface in line (Coalface I) always suffers the heaviest losses. Such an uneven distribution of losses between coalfaces will, in fact, always result from any policy where the limiting factor of the situation, in this case the conveyor belt capacity, is superimposed after the individual bunker discharge rates have been calculated. It follows, therefore, that any policy design for bunker discharge rates should also include an allocation of the available belt capacity between the bunkers.

Policy III: Conveyor Belt Allocation Policy for Bunker Discharge

This policy was designed on the basis of the points discussed in the last section. It was assumed here that if a bunker was full, it would be discharged at its maximum rate. Any residual belt capacity would then be allocated between unfilled bunkers, in proportion to the ratio of their individual levels to the sum of levels in the unfilled bunkers. A specimen calculation is given in Figure 7.5 to show the workings of this policy.

Results from using this initial version of Policy III in the model showed considerable improvements in efficiencies. However, experience with the

---

Bunker capacity = 500 tons. Maximum bunker discharge rate = 900 tons/hour. Conveyor Belt Capacity = 2000 tons/hour.

Actual Bunker Levels:	BUNKER A 500	BUNKER B 100	BUNKER C 100

Number of Bunkers Full = 1

Conveyor Belt Capacity Left = 2000 - (1 * 900) = 1100 tons/hour

Sum of levels in unfilled bunkers = 700 - (500 * 1) = 200 tons

Discharge Rate from Bunker A = Maximum Discharge Rate = 900 tons/hour

Discharge Rate from Bunker B = $\frac{100}{200}$ * 1100 = 550 tons/hour

Discharge Rate from Bunker C = $\frac{100}{200}$ * 1100 = 550 tons/hour

Total Bunker Discharge Rate = 2000 tons/hour

Figure 7.5 Worked example of Policy II

policy led to the discovery that circumstances could arise where the belt capacity allocated to a bunker could exceed its maximum discharge rate, resulting in a loss in conveyor belt utilisation. Such a situation is given in the specimen calculations of Figure 7.6, hence, a second version of Policy III was evolved to overcome this problem. This policy is specified in the diagram of Figure 7.7 and its equations are highlighted in bold type in Table 3, Appendix 3.

Bunker capacity = 500 tons. Maximum bunker discharge rate = 900 tons/hour. Conveyor Belt Capacity = 2000 tons/hour.

ACTUAL BUNKER LEVELS:	BUNKER A 400	BUNKER B 200	BUNKER C 200

Number of Bunkers Full = 0

Conveyor Belt Capacity Left = 2000 - 0 = 2000 tons/hour

Sum of levels in unfilled bunkers = 800 tons

Discharge Rate from Bunker A = Min ($\frac{400}{800}$ * 2000, 900) = 900 tons/hour

Discharge Rate from Bunker B = Min ($\frac{200}{800}$ * 2000, 900) = 500 tons/hour

Discharge Rate from Bunker C = Min ($\frac{200}{800}$ * 2000, 900) = 500 tons/hour

Total Bunker Discharge Rate = 1900 tons/hour
(Shortfall from conveyor belt capacity = 100 tons/hour)

Figure 7.6 Second worked example of Policy II demonstrating its deficiency

Having set all full bunkers to discharge at their maximum rates, the revised policy determines whether or not the allocation of the remaining conveyor belt capacity to the other bunkers will result in their maximum discharge rates being exceeded. It does this by calculating the level of each bunker at which this situation will occur. This level is referred to as the saturation level. If the bunker level exceeds this saturation level, then the discharge of that bunker is also set to its maximum rate, and any remaining conveyor belt capacity is allocated between unsaturated bunkers in the same proportion as their level to the sum of levels in all unsaturated bunkers. In the event of all three bunkers being saturated the conveyor belt capacity is allocated equally to each bunker.

The specimen calculation given in Figure 7.6 is repeated in Figure 7.8 to demonstrate how the new policy overcomes the deficiency of the preliminary version of Policy III.

The results of applying the final version of Policy III are shown in Table 7.1. It can be seen that Policy III gives an improvement in overall efficiency over Policy II on all runs. In particular, the effect of increasing the bunker capacity at low conveyor belt capacities is much more marked. For example, comparison of runs 1 and 3 for Policy II in Table 1, show a 5.57% improvement in efficiency, while comparisons of runs 1 and 3 for Policy III in Figure 4 show a 9.67% improvement in efficiency. In other words, the improved belt utilisation resulting from the improved control policy can be interpreted as a saving in bunker capacity to achieve a given efficiency.

Run 9 for Policy III also confirms the ability of this policy to reach the maximum system efficiency. Figures 7.9a and 7.9b compare the coal flow rates at each bunker discharge point with the conveyor belt capacity for both Policies II and III. The improved conveyor belt utilisation achieved by the latter can be clearly seen from these figures.

It will be noted that Run 2 now achieves a very similar efficiency to Run 1. This did not apply to Policies I and II and confirms that Policy III is, additionally, independent of the maximum bunker discharge rate.

SUMMARY AND CONCLUSIONS FROM THE POLICY ANALYSIS

The foregoing study represents an investigation into a 'hard' type of system where, despite the lack of a reference mode of behaviour, it was not difficult to conceptualise an appropriate model and to build user confidence in its ability to adequately represent the real world situation. The model created was clearly not totally endogenous, but aimed at designing control to enable the system to improve its ability to deal with a set of fluctuating exogenous inputs.

The analysis was based on testing policies against these exogenous shocks over a range of physical parameters of the system.

The level of understanding into the nature of the system which was generated by the model and experiments was sufficient to facilitate the generation of improved control policies, which were demonstrated to be capable of enhancing system performance. This is a clear example of the way in which System Dynamics facilitates policy design through understanding and how such models aid their own development.

QUANTIFYING THE BENEFITS OF ALTERNATIVE BUNKER DISCHARGE POLICIES

The foregoing results clearly indicate that it is possible to improve the overall efficiency of a conveyor system by increasing the belt and/or bunker capacities and/or by instigating more sophisticated control over bunker

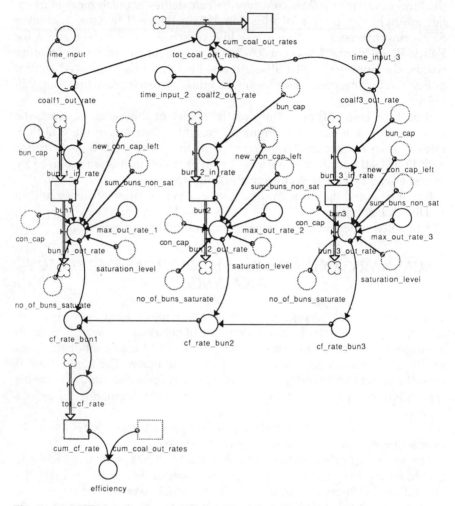

Figure 7.7 STELLA diagram for Policy III of coal clearance model

discharge rates. This leads to the interesting question as to which alternative method should be employed.

The question can be addressed by calculating how much bunker capacity would be necessary to achieve maximum efficiency, for each level of conveyor belt capacity and each bunker discharge policy used. Such figures were determined by repeating the previous simulation runs under the assumption of infinitely large bunker capacities and measuring the maximum level achieved in each bunker. These results are shown in Table

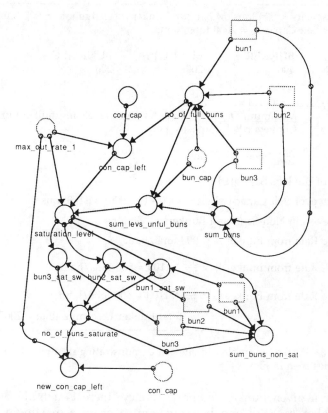

Figure 7.7 *continued* STELLA diagram for Policy III of coal clearance model

7.2 for Policies II and III, where the bunker level quoted is the sum of the maximum levels achieved in each bunker.

All runs, under Policy III resulted in the attainment of maximum system efficiency. Under Policy II, however, problems occur in determining the maximum bunker levels required. This arises because the bunker discharge policy itself interacts with the bunker level. If a large bunker capacity is introduced, the bunker discharge rate and the efficiency fall. The problem cannot be totally overcome without destroying the policy, but can be partially overcome by fixing the bunker capacity, not at an infinite value, but at a level only slightly in excess of the maximum bunker level anticipated to be needed in each run. The results for Policy II in Table 7.2 were determined in this way and, hence, represent an approximation to the total bunker capacity required.

Nevertheless, the results in Table 7.2, which are interpreted graphically (for Policies II and III) in Fig. 7.10, clearly show the following:

Bunker capacity = 500 tons. Maximum bunker discharge rate = 900 tons/hour.
Conveyor Belt Capacity = 2000 tons/hour.

ACTUAL BUNKER LEVELS:	BUNKER A 400	BUNKER B 200	BUNKER C 200

Bunker Saturation Level =
$$\frac{\text{Maximum Discharge Rate} * \text{sum of levels in unfilled bunkers}}{\text{Conveyor Belt Capacity Left}}$$

$$\frac{900}{2000} * 800 = 350 \text{ tons}$$

Number of Bunkers Saturated = 1

New Conveyor Belt Capacity Left = 2000 - (1 * 900) = 1100 tons/hour

Sum of levels in Non-Saturated Bunkers = 400 tons

Discharge Rate from Bunker A = 900 tons/hour

Discharge Rate from Bunker B = $\frac{200}{400} * 1100 = 550$ tons/hour

Discharge Rate from Bunker C = $\frac{200}{400} * 1100 = 500$ tons/hour

Total Bunker Discharge Rate = 2000 tons/hour

Figure 7.8 Worked example of Policy III demonstrating its ability to overcome the deficiencies seen in Figure 7.6

(i) For a given bunker discharge policy, there is a trade off between bunker capacity and belt capacity to achieve maximum efficiency, i.e. at low conveyor belt capacity very high bunker capacity is required, which deceases as conveyor belt size is increased. The ultimate extension of this statement is that at very high conveyor belt capacity no bunkerage is necessary, and at very low conveyor belt capacity infinite bunkerage is necessary. The consideration of these extremes is an enlightening exercise, since bunkers in conveyor systems and, indeed, stocks in any system, are often taken for granted and their primary purpose rarely considered. Additionally, bunker capacity is usually only assessed after conveyor belt capacity has been determined, rather than in conjunction with it.

(ii) Improved bunker discharge policies can reduce the physical capacity necessary to achieve maximum efficiency.

Obviously the ultimate criterion of choice between the alternative methods of achieving maximum efficiency is that of cost, and Table 7.3 presents the results of Table 7.2 in cost terms. Each conveyor belt–bunker combination, has been converted into total cost terms by summing the

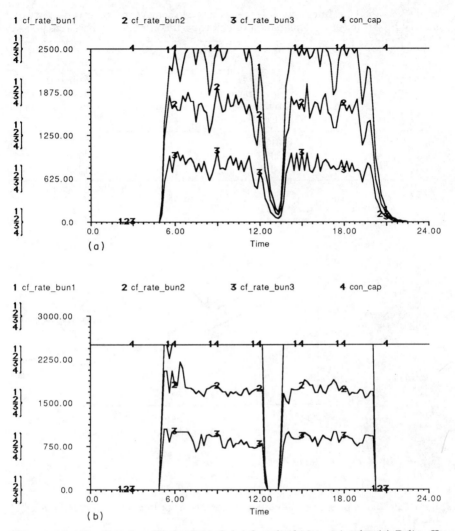

Figures 7.9 Graphs of coalflows at each bunker discharge point for (a) Policy II, and (b) Policy III (bunker capacity 500 tons, conveyor belt capacity 2500 tons/hour)

required belt capacity (at an assumed cost of £1000/ton/hour) to the required bunker capacity (at an assumed cost of capacity of £2000/ton). These are current average costs/unit of capacity and are taken as approximately representative of those necessary to uprate capacity.

It will be seen that the lowest total cost of each bunker–belt combination occurs, for any given policy, by using the largest conveyor belt size. This

Table 7.2 Total maximum buker levels required to achieve maximum efficiency (i.e. no coal losses) for each combination of belt capacity and bunker discharge policy.

Belt capacity (tons/hr)	Bunker discharge policies	
	Policy II Total maximum bunker level (tons)	Policy III Total maximum bunker level (tons)
2000	6150	5300
2500	3000	2390
3500	1854	1288

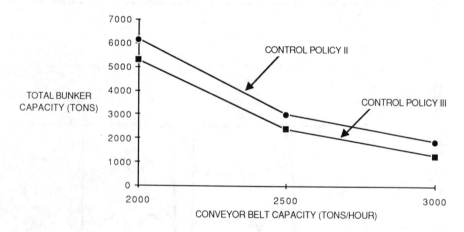

Figure 7.10 Graph of Table 7.2

Table 7.3 Costs of each belt–bunker combination for bunker discharge Policies 11 and 111.

Bunker Discharge Policy					
Policy II			Policy III		
Belt capacity (tons/hr)	Bunker capacity (tons)	Cost (£m)	Belt capacity (tons/hr)	Bunker capacity (tons)	Cost (£m)
2000	6150	14.2	2000	5300	12.6
2500	3020	8.5	2500	2390	7.3
3000	1854	6.7	3000	1288	5.5

implies that it should always be more economical to maximise belt size and minimise bunker capacity. This does, however, ignore the risk of breakdown associated with such an arrangement.

By reading across the rows, Table 7.3 also shows the order of savings associated with the type of control policy used, which are attributable to better conveyor utilisation and hence, lower bunkerage requirements. In all cases these will be seen to be very substantial. For example the effect of using Policy III rather than II at a belt capacity of 2500 tons/hour results in a capital saving of £1.2m (less of course the cost of control).

TRANSFER OF THE ALGORITHM TO OTHER SYSTEMS

The original objective for the coal clearance study was to improve managerial understanding of feedback control and its implications, particularly as an alternative to capacity expansion, in improving the performance of such systems. The rationale for using simulation analysis was chiefly because the ideas of control could be developed in an understandable way and easily communicated by this medium.

It quickly became apparent during this study that such a medium could be usefully applied to communicating the same ideas between different types of system. The objective of the remainder of this chapter is, therefore, to explore the versatility and isomorphic properties of the developed control policy (or algorithm). Its potential use as a means of conceptualising models in other systems and its generalisation will both be explored using influence diagrams.

It should be noted that no attempt is made to present the final version of Policy III in influence diagram form. This is possible, but it is felt that the complexity of the resultant diagram is not consistent with the above objective of ease of communication and further attention here will be limited to the initial version of Policy III. An influence diagram of this is shown for a two bunker system in Figure 7.11.

It will be seen from Figure 7.11 that each bunker level is controlled by a simple negative feedback loop (loop 1). When a bunker is not full, but levels in other bunkers are higher than its own, loop 1 for that bunker is dominated by the positive loop (2). When a bunker is full, a reinforcement of the basic control loop for that bunker (loop 3) takes place and this weakens loops 2 and 4 for other unfilled bunkers and hence weakens control of them, however, loop 3 is undermined by loop 4, if other bunkers are also full.

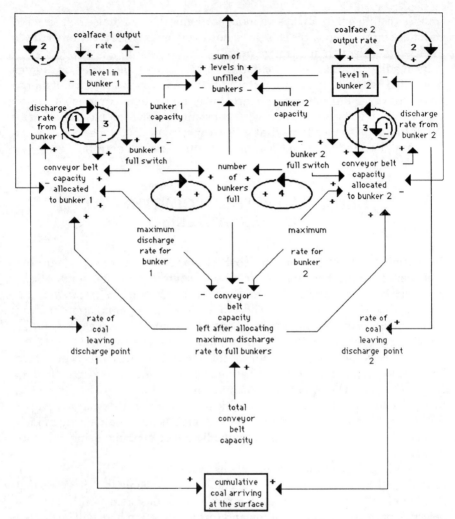

Figure 7.11 Influence diagram of coal clearance model incorporating bunker discharge Policy III

PRODUCTION POLICY DESIGN IN COAL MINES

The first system to which the developed control algorithm was transferred was that of a whole coal mine. A study was undertaken to analyse the management control of production operations within a coal mine in the light of new information retrieval technology (Holmes 1980, Wolstenholme and Holmes 1985). The aim of the study was to design management policies for improving the overall level and stability of total mine output in the light of geological uncertainty.

The process of coal mining can be described as a set of sequential operations. These involve the preparation or development of coal panels ready for production and producing from them by extracting the coal in strips from the face of the panel (longwall mining). Since each mine normally has a number of such production coalfaces (and associated development panels), the major managerial problem is essentially one of how to regulate output between them. This situation is, of course, entirely analogous to that encountered in the coal clearance model and, hence, it is possible to establish variables for a model at the total mine level analogous to each of the variables in Figure 7.11.

A diagram containing such variables is presented in Figure 7.12. The construction of this diagram will now be described and differences from the structure of Figure 7.11 highlighted.

The state variables of this system, equivalent to the bunkers of the coal clearance model, are those of the stocks of coal contained in each panel to be mined. These must be replenished by developing new coal faces periodically. However, in the short term they are depleted by a continuously applied production rate which is equivalent to the bunker discharge rate but, of course, subject to uncontrolled variations such as geological hazards. The production rate, at a given level of productivity, is a function of the manpower or manshifts used on each face and these latter variables represent the resources which must be regulated or allocated between coalfaces. A basic mechanism for doing this can be postulated in exactly the same form as for the coal clearance model. However, to cope with the additional complexity and with existing conventions some modifications are necessary.

Firstly, the convention in mine management is to measure progress by monitoring the cumulative coal cut from the mine over a period of time rather than that remaining underground in the face stock. A planned cumulative output over a given period of time is defined, which is based on planned production rates. The way of applying control to this system as suggested by the algorithm is to monitor each coalface in terms of the discrepancy between the planned and actual cumulative coalface output rates. This procedure is shown in Figure 7.12.

Secondly, for the purpose of illustration, manshifts and manpower resources are shown together and, thirdly, the concept of a variable resource size is introduced. This is because, in practice, the upper limit of both manshifts and manpower available can be controlled, but are also subject to uncertainty.

The bunker full switch of Figure 7.11 is replaced in Figure 7.12 by a coalface criticality switch, based on the size of the cumulative production

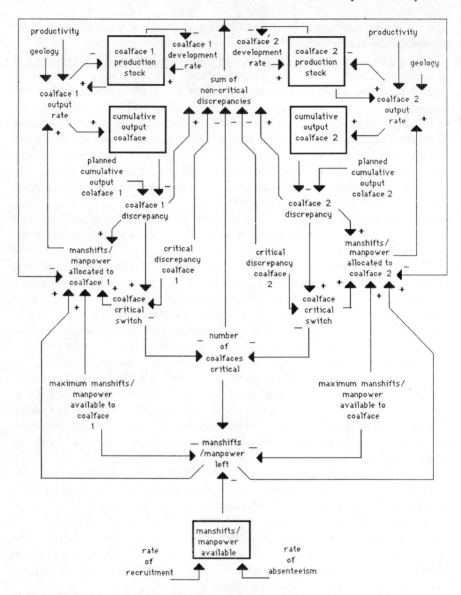

Figure 7.12 Influence diagram of the coal mine model

discrepancy for each coalface, relative to a critical size of discrepancy. Application of the algorithm then dictates that allocation of the maximum allowable manshifts or manpower is made to each coalface whenever the cumulative output discrepancy is critical. Otherwise, the allocation is made in proportion to the size of the individual coalface discrepancy relative to the sum of discrepancies on non-critical coalfaces (equivalent to the sum of levels in unfilled bunkers in Figure 7.11).

This control mechanism depicted in Figure 7.12 is hence exactly the same as shown in Figure 7.11 and similar feedback loops exist. It should be noted, however, that control is instigated in terms of discrepancies in cumulative output, rather than discrepancies in the stock which is depleted. This means that the signs of some of the links constituting the major control loops are changed. The overall loop polarities remain the same, however.

The transfer of the control algorithm with its minor modifications has immediate implications for the new system. In particular, it implies that certain variables are important and suggests the way in which these should be used. The most interesting of these variables are the critical discrepancies for each coalface, the maximum manshifts or manpower which can be allocated, and the relationships needed between manshifts/manpower and output.

In practice, coalfaces are usually defined as being behind or ahead of schedule and, whilst every effort is made to correct discrepancies no specific mechanism for this exists. As discrepancies increase there is no physical barrier to be encountered, equivalent to the bunker capacity of Figure 7.11, which shuts the system down. Rather, a gradual deterioration of total output takes place with a resultant mis-phasing of coalface finish times and problems in creating resources for future developments. There is, in fact, always a temptation to change the planned output figures when discrepancies become large, rather than to instigate control.

Consequently, the concept of a critical discrepancy captures a very important issue for mine management, focuses attention on it and leads to discussion of the factors determining the point at which control action should be instigated.

The need to define the maximum size of the resource allocation feasible for each coalface also raises interesting insights. This is equivalent to defining the maximum bunker discharge rate for the coal clearance model. Thought must be given as to whether coalface output can be continuously controlled in response to management inputs or if this must be simply switched off when insufficient manpower exists, and on when sufficient exists. Attention is further directed by the diagram to the issue of whether manpower can be transferred from other tasks (for example the coalface development rate) and how much flexibility exists for changing shifting patterns.

ALLOCATION OF DEFENCE RESOURCES

Figure 7.13 depicts the application of the algorithm to a model of a defence situation under threat from two directions (axes), which immediately generates a unique perspective of this system and the implications of control for it. This situation is depicted as directly analogous in structure to the coal clearance model, where the size of the cumulative threat (bunker level) is measured and reduced by a threat elimination rate (bunker discharge rate).

The variables denoted as 'cumulative threat' represent a generalised notion concerning how the defenders feel about the security of their

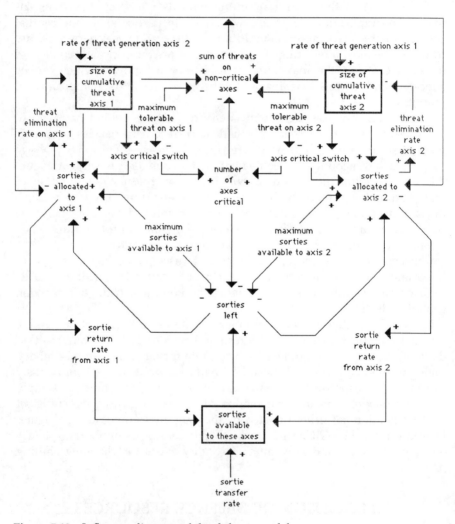

Figure 7.13 Influence diagram of the defence model

situation on each axis. This may reflect a number of factors in practice, such as their perception of the size of the enemy force and the sophistication of, and their confidence in, the available surveillance equipment. Such factors, together with knowledge of the situation on other axes, may provide a basis for determining the maximum tolerable threat level and, hence, define the critical point at which escalation of action is called for. Action of this type will involve the deployment of the defender's own forces and involve the allocation of people, armament or other machines to each axis.

In Figure 7.13 it is assumed that a number of sorties are pre-allocated and that the problem is one of how to sub-allocate sorties between the two axes, in order to delay the enemy advance and decrease the cumulative threat. Again, thought must be given to the definition of the maximum sorties available to each axis (perhaps in order to keep some spare or for priority call to other axes) and also to the minimum sorties considered necessary to present evidence of the defendants existence and capability.

GENERALISATION OF THE ALGORITHM

The foregoing examples have demonstrated the application of the derived control algorithm in a number of different systems. A generalisation of the algorithm can now be presented in the light of the experience gained and this is shown in the influence diagram of Figure 7.14.

The resource to be allocated is designated as the controlling resource for the system, which is to be allocated between a number of competing processes, two of which are represented in Figure 7.14. The rate of allocation in each process generates a rate of conversion, depending on the productivity of the controlling resource in that process. The resource conversion rate can also, of course, be subject to exogenous shocks which, by definition are outside the control of the system. The rate of conversion of the controlled resource in each process transforms this resource from a prior to a post state.

Either the prior or post state of the processes can be used for control in a given system. Figure 7.14 shows the use of the post state, for this purpose, which is compared with its planned version and any discrepancy is, in turn, compared with a pre-defined critical discrepancy value. The rate of allocation of the controlling resource to each process is then made at a defined maximum feasible rate, if the discrepancy is critical for any process. Otherwise, it is made in proportion to the ratio of the discrepancy to the total discrepancies associated with the non-critical post states. The resultant rate of conversion of the controlled resource in each process then influences the size of the controlling resource available.

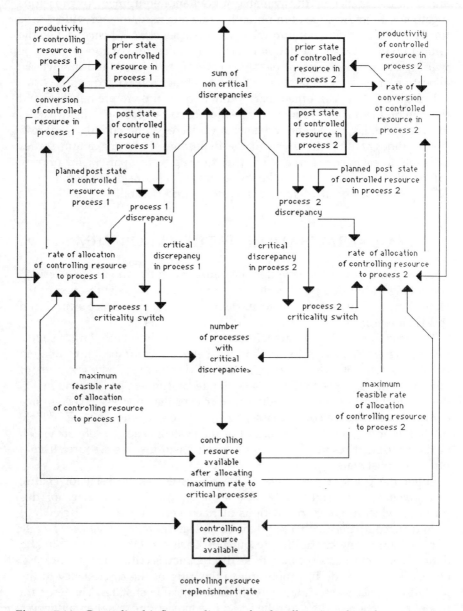

Figure 7.14   Generalised influence diagram for the allocation algorithm

COMMENTS ON THE USE OF PORTABLE CONTROL ALGORITHMS

This chapter has demonstrated that generic control structures can be developed which are applicable in a wide range of systems, and that the application of these structures can help in both conceptualising models and in improving control in the recipient systems.

The specific interpretation of one such structure for the control of allocation has been shown to generate interesting insight and understanding in a number of very different systems. It is felt that the definition of the generalised influence diagram of this control structure highlights the portability of the concept in general, and presents a formalised approach to the process of inter-system transfer of the allocation algorithm in particular.

One of the major problems in using portable algorithms, particularly in the management context, is that of creating structures which are simple and general enough to transfer between systems and yet sophisticated and specific enough to generate new insight in the recipient system. There is obviously a danger of transferring too complex a structure which can lead to inappropriate control design.

The structures must, at a fundamental level, be independent of the physical systems to which they are applied and yet capable of matching the reality of the environment into which they are placed.

Whilst it is important to recognise that the transfer of an algorithm between systems, can itself change thinking by transferring the technology of the donor system to that of the recipient system, it must also be appreciated that a tailoring process is usually required. This might mean relaxing some of the assumptions contained in the algorithm, which might ultimately undermine the advantage gained.

REFERENCES

Holmes, R.K. (1982) *The Design of Colliery Information and Control Systems*, Doctoral Thesis, University of Bradford Management Centre.

Wolstenholme, E.F. and R.G. Coyle, (1980) *The Incorporation of Discrete Factors in System Dynamics Models*, DYNAMICA No. 6.

Wolstenholme, E.F. and R.K. Holmes, (1985) The Design of Colliery Information and Control Systems, *European Journal of Operational Research*, **20**, 168–181.

Chapter 8

A Case Study in Defence Analysis

INTRODUCTION

This chapter describes the application of system dynamics to the analysis of an army defence situation and demonstrates the types of insight which can be generated.

The approach used is somewhat different from current defence modelling practice. Most traditional modelling of ground conflict has centred on assessing the actual outcome resulting from the face to face confrontation of two combatants. Such low aggregation, high resolution modelling is not considered in the approach here. Instead, in line with general developments in defence modelling (Huber 1985), emphasis is placed upon assessing the use of indirect strategies aimed at avoiding situations occurring where face to face confrontation will lead to a defeat. That is, in the defender's terms, the aim is to design indirect strategies which will result in reducing those characteristics of the attacker (such as speed and force size) to an acceptable level on arrival at the attacker's position. Conversely, in the attacker's terms, the aim is to design strategies which will result in preserving such characteristics.

The model developed in this chapter is considerably larger than those encountered so far. One of the major problems with developing computer models of complex systems is that the models themselves become more and more complex. This frequently results in them assuming a life and purpose of their own, quite separate from their role in facilitating understanding of the system which they represent. This tendency is minimised in System Dynamics by maintaining, in parallel with the complex simulation model, a simplified diagrammatic, qualitative model of the underlying feedback structure of the quantitative model. This might be an updated version of the original diagram from which the quantitative model was developed or an abstracted version of it. The important point is that the two models must be developed together and it is only when the quantitative model development rate outstrips the qualitative that confusion arises. The qualitative model promotes a means to interpret the results of the quantitative model to create a lasting explanation of the operation of the system and a basis for further qualitative or quantitative analysis. The cycle of model development suggested by these procedures is shown in Figure 8.1.

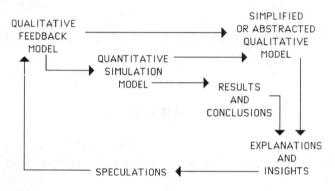

Figure 8.1 The process of parallel development of qualitative and quantitative models

The uses of such ideas are, in practice, quite subtle and one of the purposes of this chapter is to demonstrate their application. Firstly, the purpose of the study will be presented and both a qualitative and quantitative model (referred to as the Armoured Advance Model) developed for analysis of the problem described. This is followed by a presentation of the experiments devised to analyse the model and some specific results and conclusions. This approach follows the conventional route through the inner loop of Figure 8.1. Secondly, an abstract qualitative model will be developed to assist the interpretation and generalisation of the results, that is, following the outer loop in Figure 8.1.

THE PROBLEM

The specific problem addressed was how to investigate the merits of alternative strategies by a static, defensive force (Blue) for slowing down the advance of an attacking force (Red), under a number of adaptive strategies by the latter concerning the timing of its formation changes. General defence thinking on this issue suggests that basically Red's alternatives are to change to a more widely dispersed (company column) formation early in the advance, in order to reduce its vulnerability to attack, or to maintain a dense (battalion) formation for as long as possible, since a higher speed is attainable. However, it is unclear as to which of these alternatives is better and how Blue's defensive strategies, concerning its delivery of fire, might interact with them.

This is a diffuse, ambiguous and subjective question, in that vulnerability and dispersion are concepts which are difficult to define and quantify. It is also complicated by containing both spatial and time dimensions.

In the absence of a real world system from which observations for a reference mode of behaviour could be made, a system dynamics model was developed using the modular approach to conceptualisation.

THE IDENTIFICATION OF RESOURCE STATES AND RESOURCE FLOWS

Figure 8.2 lists the resources identified as being relevant to the problem described. The first two of these are the number of Red units advancing and the speed of these units, since it is anticipated that Blue fire will have the effect of both attritting Red numbers and reducing Red speed. The third is Red distance, since this will change as the gap between the combatants closes during the advance, and will affect the rate at which Blue can inflict attrition and speed loss on Red. The fourth is Blue ammunition, since this will significantly affect the rate of fire delivery by Blue.

There are two states to be considered for each of the first three resources listed in Figure 8.2. These are the number of Red units, the Red speed and the Red distance in both battalion and company column formations. There is only one state relevant for the resource of Blue ammunition, which is the total stock of this available.

Figure 8.2 also shows the resource flows constructed for each resource. The two states of Red units are cascaded together, since the rate of deployment from battalion formation to company columns, transfers units between the two states. The advance start rate feeds the number of units in battalion formation and the rate of arrival at the Blue position depletes the number of units in company columns. Additionally, both states are subject to loss by attrition.

The structure of the resource flows for Red speed in each formation are identical. In both cases speed can be lost when Blue is delivering fire and recovered when it is not.

The structure of the resource flows for Red distance in each formation are also identical. Since only advance is considered for Red, the distance travelled by Red increases as the rate of change of distance increases. The rate of change of distance is, of course, equivalent to the speed of Red, which provides a direct link between these two resources.

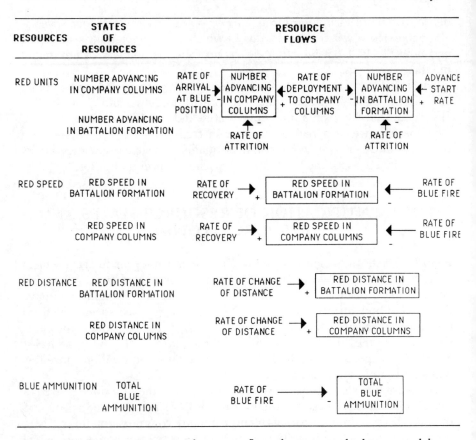

Figure 8.2 Resources, states and resource flows for armoured advance model

In the absence of ammunition resupply, the resource flow associated with Blue ammunition consists of the total ammunition stock and its rate of depletion (Blue weapon delivery rate).

THE COMPOSITE MODEL

Figure 8.3 shows an outline influence diagram for the composite model. The actual movement of Red's advance can be traced by the variables across the top of the diagram. Units for the advance are assembled in a pre-defined area and advance takes place firstly, in a dense 'battalion' formation and secondly in a slower, but more dispersed, 'company column' formation.

In practice Red would also undergo perhaps a further change in formation

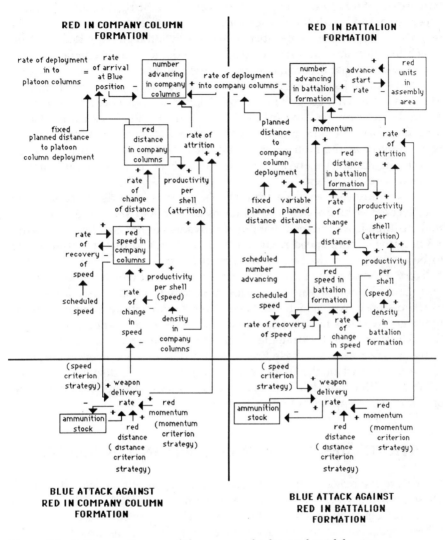

Figure 8.3 Influence diagram of the armoured advanced model

to platoon columns. Only two formations are considered in the model described here and the rate of change to platoon columns shown in Figure 8.3 is effectively the Red arrival rate at the Blue position.

The key strategy variables for Red in this chain are the rates of deployment between formations, which are considered to take place at a fixed distance (preplanned response) or a variable distance (adaptive response). The variable distance strategy represents an attempt by Red to either delay formation change to take advantage of the higher scheduled speed associated with the denser formations or to change formation early to preserve force size. A number of secondary strategies for Red exist by which to determine the variable distances to the formation change points.

It is assumed that Red will base the decision of the timing of the formation change point on its own speed of advance and numbers of units advancing. The formation change point will be delayed as the speed falls behind schedule and brought forward as the number of units advancing falls, as a result of Blue fire. The product of the speed and numbers advancing can be thought of as the momentum of advance. Momentum is an often used concept in military analysis, but has rarely been used as a quantitative measure as defined here.

The key strategic variable from Blue's point of view is, of course, the effectiveness of its fire delivery, in terms of both Red speed reduction and Red attrition. It will be seen that the effectiveness of fire is defined in Figure 8.3 in terms of both speed reduction and attrition and that it is itself made to be a function of the rate of weapon delivery and the productivity per delivery. Productivity of fire is an interesting concept which is analogous to managerial labour productivity. This productivity is obviously a function of the distance over which fire takes place (accuracy) and the density of the target, which is in turn a function of the type of formation assumed by Red. There are various strategies available to Blue for the delivery of fire. Three possibilities are to base fire on Red distance, speed or momentum.

The lower part of the diagram in Figure 8.3, which displays variables relating to Blue fire delivery, speed and distance, are replicated for the battalion and company column situations.

STRATEGIES AND PARAMETERS IN THE MODEL

To analyse the major strategy options, several experiments were designed. These involved setting parameter levels within the model.

The first set of parameters to be defined relate to Blue's strategy options. When Blue delivers fire on a distance criterion, it is assumed that this delivery is only over a proportion of Red's distance of advance in both battalion and company column formations. The proportions defined were the initial 20%, and that lying between 80% and 90% of the distance of advance in battalion formation. In company columns the corresponding figures were 30% and 70% to 90% of Red's distance of advance. When Blue fire was switched on the fire was delivered at one of two predefined (fixed) rates. These were light fire (700 shells/hour) or heavy fire (1500 shells/hour).

In the case of Blue delivering fire on a speed criterion, it is assumed that Blue's fire delivery is switched on when Red's speed exceeds a preset upper limit and switched off when Red's speed is driven below a preset lower limit in both formations. These limits were set to 30 and 18 kmh^{-1} respectively in battalion formation and 12 and 8 kmh^{-1} respectively in company columns. The maximum speeds achievable were set to 40 kmh^{-1} in battalion formation and 15 kmh^{-1} in company columns. Again fire was delivered at either the light or heavy rate.

In the case of Blue delivering fire on a momentum criterion, it was assumed that Blue's fire delivery was similarly switched on and off on preset upper and lower limits of Red's momentum in both formations. These limits were 30 000 (vehicles* (k/h)) and 25 000 (vehicles*k/h)) respectively in battalion formation and 12 000 (vehicles*k/h)) and 6000 (vehicles*k/h)) respectively in company columns. Again, Blue fire was delivered at either a light or a heavy rate.

It is assumed in all fire delivery strategies that Red can recover speed when Blue firing stops.

The second set of parameters to be defined are those related to Red's strategy options. These are either to change formation at a fixed distance of advance or at a variable distance of advance. The combatants are initially stationed 240 km apart and the fixed distance to formation change is specified as 160 km from the beginning of the advance. It is assumed in the *variable* distance formation change strategy that the distance to the formation change point is a function of the normal (*fixed*) distance and two multipliers. One is dependent on Red size and one is dependent on Red speed, where the influence of the two is weighted by a speed/size parameter (SZS) as in the following equation:

$$
\begin{array}{l}
\text{Variable planned} \\
\text{distance to} \\
\text{formation} \\
\text{change point}
\end{array}
=
\begin{array}{l}
\text{Normal} \\
\text{planned distance} \\
\text{(fixed distance)}
\end{array}
*
\left[
\begin{array}{l}
\text{Red} \\
\text{size} \\
\text{multiplier}
\end{array}
*(1\text{-SZS}) +
\begin{array}{l}
\text{Red} \\
\text{speed} \\
\text{multiplier}
\end{array}
* (\text{SZS})
\right]
\quad (8.1)
$$

In all the experiments in this chapter SZS was set to 0.5 and the shape of the multipliers are shown in Figures 8.4 and 8.5. The initial size of the Red force was set to 1800 vehicles.

A third parameter of interest is that of Blue's accuracy of fire which was assumed to improve as the distance between the combatants decreased. A blue weapon accuracy multiplier was defined to represent this as shown in Figure 8.6.

The above description of the model is sufficient to facilitate an understanding of the results from experiments with the model. Readers not interested in the detail of equation formulation should move directly to the results section.

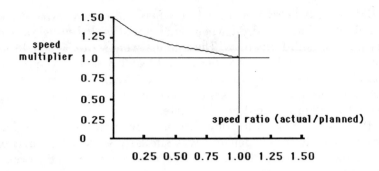

Figure 8.4   Red speed multiplier

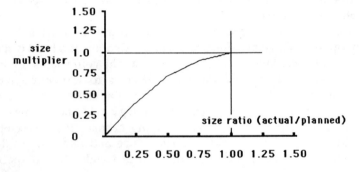

Figure 8.5   Red size mutliplier

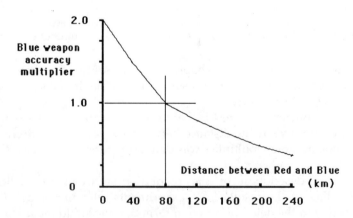

Figure 8.6   Blue weapon accuracy multiplier

Readers who are interested in the detailed formulation of equations for the model should examine the full listing of the model equations, presented in Table 1 of Appendix 4. In addition to the usual documentation, further notes of explanation are given in Table 2, Appendix 4.

The next section contains some examples of the type of equation formulation used in the model.

EQUATION FORMULATION

A detailed description of the equations for four sections of the model is presented here. These relate to the Red movements in battalion formation, the Red formation change from battalion formation to company columns, the Blue fire delivery and the calculation of Red arrival time at the Blue position. The equations for these activities are given in Tables 8.1, 8.2, 8.3 and 8.4 respectively.

Equations for the Red Movement in Battalion Formation

Consider Table 8.1, which contains equations 8.1.1 to 8.1.7. Equation 8.1.2 is a level equation describing the number of Red units in the assembly area, ready to advance. The initial number of units assembled (UIAXX=1800) is described in equations 8.1.3 and 8.1.4. The advance start rate (ASR), given in equation 8.1.1, moves all the units instantaneously, that is in one simulation interval (DT), from the assembly area to the state of advancing in battalion formation. As soon as the advance has started this rate is zero. The number of units advancing in battalion formation is described by a second-level equation (equation 8.1.6). This is increased by the advance start rate. It is depleted by the company column deployment rate (CCDR) and by the attrition rate in battalion formation (ARBF). Initially, there are no units in battalion formation (see equation 8.1.7).

The company column deployment rate equation 8.1.5 acts like the advance start rate and moves all Red units instantaneously from a battalion formation to company columns. However, this rate can only be activated when the actual distance of advance in battalion formation (ADBF) exceeds the planned distance to company column deployment (PDCCD), hence, a CLIP function is incorporated in equation 8.1.5 to impose this condition.

Red's movement in company columns is formulated in a similar way.

Equations for the Red Formation Change

Table 8.2 contains the equations for Red's formation change calculations. Equation 8.2.1. defines a constant P1. This can only take the values of zero

Table 8.1 DYSMAP2 equations for Red Movement in Battalion Formation

Equation	Equation Numbers
R ASR.KL=MAX(0,UIAA.K/DT)	(8.1.1)
L UIAA.K=UIAA.J+DT*(-ASR.JK)	(8.1.2)
N UIAA=UIAAX	(8.1.3)
C UIAAX=1800	(8.1.4)
R CCDR.KL=MAX(0,NABF.K/DT)*	(8.1.5)
X CLIP(1,0,ADBF.K,PDCCD.K)	
L NABF.K=NABF.J+DT*(ASR.JK-CCDR.JK-ARBF.JK)	(8.1.6)
N NABF=0	(8.1.7)

```
DOCUMENTATION (V=VEHICLES, H=HOURS, K=KILOMETRES)

D ADBF=(K). . . .   ACTUAL DISTANCE IN BATTALION FORMATION
D ARBF=(V/H). . .   ATTRITION RATE IN BATTALION FORMATION
D ASR=(V/H) . . .   ADVANCE START RATE
D CCDR=(V/H). . .   COMPANY COLUMN DEPLOYMENT RATE
D NABF=(V). . . .   NUMBER ADVANCING IN BATTALION FORMATION
D PDCCD=(K) . . .   PLANNED DISTANCE TO COMPANY COLUMN FORMATION
D UIAA=(V). . . .   UNITS IN ASSEMBLY AREA
D UIAAX=(V) . . .   INITIAL UNITS IN ASSEMBLY AREA
```

or 1. When it is equal to zero the formation change from battalion formation to company columns will take place at a fixed distance. When it is equal to 1 the formation change will take place at a variable distance, defined by other equations, hence, P1 is a strategy switch for Red.

The planned distance to company column formation (PDCCD), defined in equation 8.2.2 can, therefore, take on one of two values, depending on the switch P1; the fixed distance to company column deployment (FDCCD), which is 160 km (see equation 8.2.3), or the variable distance (VDCCD).

Equation 8.2.4 has already been encountered in the previous section of this chapter on strategies and parameters for the model. This equation describes the variable distance to company column deployment, which is the product of one of two possible multipliers and the fixed distance. These multipliers are the battalion formation speed (BFSM) and size (BFZM) multipliers, already described in Figures 8.4 and 8.5, respectively. The weight attached to each of these multipliers is determined by the speed/size parameter (SZS). When this is equal to 1, the variable distance to company column deployment is determined by the speed multiplier. Conversely, when SZS is equal to zero, the same distance is determined solely by the size multiplier. As will be seen in equation 8.2.6, SZS was set to a value of 0.5 for all experiments presented in this chapter.

In addition, the formation change is constrained to take place at one kilometre from the total distance of advance (TD-1), if it has not already taken place. This ensures that Red always arrives at the Blue position in a

Table 8.2 DYSMAP2 equations for Red formation change calculations

Equations	Equation Numbers
C P1=0	(8.2.1)
A PDCCD.K=P1*FDCCD+(1-P1)*VDCCD.K	(8.2.2)
C FDCCD=160	(8.3.3)
A VDCCD.K=(MIN(FDCCD*BFSM.K,TD-1)*SZS	(8.2.4)
X +MIN(FDCCD*BFZM.K,TD-1)*(1-SZS)	
C TD=240	(8.2.5)
C SZS=0.5	(8.2.6)
A BFZM.K=TABHL(BFZMT,BFZR.K,0,1.0,.20)	(8.2.7)
T BFZMT=0/.55/.75/.90/.97/1.0	(8.2.8)
A BFZR.K=(NABF.K/UIAAX)*BFS.K	(8.2.9)
A BFSM.K=TABHL(BFSMT,BFSR.K,0,1.0,.25)	(8.2.10)
T BFSMT=1.5/1.28/1.18/1.09/1.0	(8.2.11)
C BFSR.K=(ASBF.K/PSBF)*BFS.K	(8.2.12)
A AMBF.K=ASBF.K*NABF.K	(8.2.13)
C PSBF=40	(8.2.14)

DOCUMENTATION (V=VEHICLES, H=HOURS, K=KILOMETRES)

```
D AMBF=(V*(K/H))  ACTUAL MOMENTUM IN BATTALION FORMATION
D BFSM=(1)        BATTALION FORMATION SPEED MULTIPLIER
D BFSMT=(1)       BATTALION FORMATION SPEED TABLE
D BFSR(1)         BATTALION FORMATION SPEED RATIO
D BFZM=(1)        BATTALION FORMATION SIZE MULTIPLIER
D BFZMT=(1)       BATTALION FORMATION SIZE TABLE
D BFZR=(1)        BATTALION FORMATION SIZE RATIO
D FDCCD=(K)       FIXED DISTANCE TO COMPANY COLUMN DEPLOYMENT
D PSBF=(K/H)      PLANNED SPEED IN BATTALION FORMATION
D PDCCD=(K)       PLANNED DISTANCE TO COMPANY COLUMN DEPLOYMENT
D P1=(1)          COMPANY COLUMN POLICY SWITCH
D SZS=(1)         SPEED/SIZE RATIO SWITCH (IF =1 SPEED RATIO
X                 USED IF =0 SIZE RATIO USED)
D TD=(K)          TOTAL DISTANCE
D UIAAX=(V)       INITIAL UNITS IN ASSEMBLY AREA
D VDCCD=(K)       VARIABLE DISTANCE TO COMPANY COLUMN
X                    DEPLOYMENT
```

company column formation. The total distance of the advance is defined as 240 km (see equation 8.2.5).

The equations for the battalion formations size and speed multiplier equations are given in equations 8.2.7, 8.2.8 and 8.2.10, 8.2.11, respectively and are constructed using table functions. The ratios for the x-axes of these tables are formulated in equations 8.2.9 and 8.2.12, respectively. The size ratio consists of the number of units advancing in battalion formation over the numbers of units initially available in the assembly area. The speed ratio consists of the actual speed of Red in battalion formation over the planned speed for Red in battalion formation. The latter is 40 kmh[1] as defined in equation 8.2.14. Equation 8.2.13 calculates the actual momentum of Red in battalion formation as the product of its speed and size.

Equations for Blue Fire Delivery

Table 8.3 contains the equations for the fire delivery by Blue on a speed criteria when Red is in battalion formation. (Fire delivered by Blue against Red in battalion formation is assumed to be by air strike and against company columns by artillery (guns)).

Table 8.3 DYSMAP2 equations for Blue fire delivery on a speed criteria against Red in battalion formation

Equation	Equation Numbers
A BFS.K=CLIP(1,0,NABF.K,1)	(8.3.1)
A WDRAS.K=PWDRA*BFS.K*FLSB.K	(8.3.2)
C PWDRA=1500	(8.3.3)
L OSBF.K=OSBF.J+DT*(ASBF.J-OSBF.J)/DT	(8.3.4)
N OSBF=0	(8.3.5)
A FLSB.K=CLIP(1,0,ASBF.K,BFSLL)	(8.3.6)
X *CLIP(1,CLIP(0,1,ASBF.K,OSBF.K),ASBF.K,BFSUL)	
C BFSLL=1	(8.3.7)
C BFSUL=30	(8.3.8)

DOCUMENTATION (K=KILOMETRES, H=HOURS, S=SHELLS, V=VEHICLES)

D	ASBF=(V/H)	ACTUAL SPEED IN BATTALION FORMATION
D	BFSLL=(K/H)	BATTALION FORMATION SPEED LOWER LIMIT
D	BFSUL=(K/H)	BATTALION FORMATION SPEED UPPER LIMIT
D	BFS=(1)	BATTALION FORMATION SWITCH
D	FLSB=(1)	FLIP FUNCTION FOR FIRE DELIVERY ON BATTALION
X		FORMATION ON SPEED CRITERION
D	OSBF=(K/H)	OLD VALUE OF SPEED IN BATTALION FORMATION
D	NABF=(V)	NUMBER ADVANCING IN BATTALION FORMATION
D	PWDRA=(S/H)	PLANNED WEAPON DELIVERY RATE AIRCRAFT
D	WDRAS=(S/H)	WEAPON DELIVERY RATE BY AIRCRAFT ON SPEED
X		CRITERION

First of all a battalion formation switch (BFS) is calculated in equation 8.3.1. This assumes a value of one, as long as there are greater than or equal to 1 Red unit advancing in battalion formation.

The Blue fire on a speed criterion is given by equation 8.3.2. This is the weapon delivery rate by aircraft using a speed criterion (WDRAS). It is the product of the planned weapon delivery rate, the battalion formation switch and a special function (FLSB). The planned weapon delivery rate is defined as 1500 rounds (shells)/hour in equation 8.3.3. This value denotes heavy fire, in contrast to a value of 700 rounds/hour, which denotes light fire. The battalion formation switch is incorporated so that fire against Red in battalion formation only takes place when there are Red units advancing in battalion formation.

The variable FLSB refers to a flip function for fire delivery on a speed criterion against Red in battalion formation. It is formulated in equation 8.3.6 as three CLIP functions, with the third nested within the second. FLSB is a zero–one switch which turns the firing rate on and off when different conditions apply. Two values of Red speed are involved in these conditions. These are the battalion formation speed lower limit (BSFSLL), defined in equation 8.3.7 and upper limit (BFSUL), defined in equation 8.3.9.

(A FLIP function is available in DYSMAP2 which performs the task of equation 8.3.6. The full formulation was created here to enable this model to be run under the original DYSMAP software.)

If Red's actual speed in battalion formation exceeds the defined upper limit, then both the first and second CLIP functions are 1 and fire is delivered. Firing then continues until the Red speed is driven down to the level of the lower limit, when the first function becomes zero, and fire is switched off. As speed recovers the lower limit can then be exceeded without provoking fire since, although the first CLIP has a value of 1, the second CLIP remains at a value of zero until the upper limit is again reached.

This situation, where fire is permitted on falling values of speed, but not on rising values, is achieved by the third CLIP function. This tests whether the current value of speed is greater than or equal to the value of speed one DT ago, which is defined as the old value of speed (OSBF). If it is, then speed is rising and firing not permitted. The old value of speed is calculated in a similar way to the old values in the CIR model of Chapter 6, that is, as a level, which is updated by the difference between the new value of speed and the old value in one simulation interval (DT), every simulation interval. Since, the old value is calculated here every DT, the pulse function can be omitted. The initial condition of this level is zero (see equation 8.3.5).

Equations for the Time of Arrival of Red at the Blue Position

Table 8.4 contains an example of the type of equations used to calculate the value of a variable at a particular event time in the model. For example, it is required to know the time at which Red arrives at the Blue position, the size of Red on arrival, the time to company column deployment and the size of the Red force at company column deployment. These can all be calculated in a similar way. The specific example given in Table 8.4 is for the time at which Red arrives at the Blue position.

Table 8.4 DYSMAP2 equations for the calculation of the time at which Red arrives at the Blue position

Equation	Equation Number
A ARS.K=CLIP(1,0,CD.K,TD)	(8.4.1)
L OARS.K=OARS.J+DT*((ARS.J-OARS.J)/DT)	(8.4.2)
N OARS=0	(8.4.3)
L ART.K=ART.J+DT*ARS.J*(1-OARS.J)*(TIME.J/DT)	(8.4.4)
N ART=0	(8.4.5)

DOCUMENTATION (H=HOURS)

D OARS=(1)	OLD VALUE OF RED ARRIVAL SWITCH
D ARS=(1)	SWITCH TO MARK ARRIVAL OF RED AT BLUE'S POSITION
D ART=(H)	TIME OF RED ARRIVAL AT BLUE POSITION
D TIME=(H)	SIMULATION TIME
D CD=(K)	CUMULATIVE DISTANCE

Firstly an arrival switch (ARS) is calculated (equation 8.4.1), which assumes a value of one when the cumulative distance of the Red advance (CD) equates with the total distance defined for the advance (TD). Secondly, an old value of the arrival switch (OARS) is calculated every DT (equations 8.4.2 and 8.4.3), as in Table 8.3. The time of arrival is then composed of the unique event when the arrival switch is 1, but the old arrival switch is still zero. That is, when the product of ARS and (1-OARS) is equal to 1. The arrival time (ART) is defined as a level in equation 8.4.4, which is zero (equation 8.4.5) until arrival takes place, at which point the current value of the cumulative advance time (TIME) is recorded.

RESULTS FROM THE MODEL

Some examples of results from the model are contained in Tables 8.5a and 8.5b. Figure 8.5a relates to Red's fixed distance formation change strategy and Figure 8.5b to Red's variable distance formation change strategy. Within each of these, three Blue fire strategies are listed at two levels of fire delivery (light and heavy).

As in the coal clearance model of Chapter 5, the behaviour of this model was summarised using performance measures. The performance measures used were:

- the time to company column deployment
- the total time for Red to reach Blue
- the size of the Red force on arrival at Blue's position

Table 8.5(a) Results from armoured advance model: Red using a fixed-distance strategy for formation change

Blue Strategy	Red: fixed-distance strategy	Time to company column deployment (TTCCD)	Time for Red to reach Blue (ART)	Size of Red force arriving at Blue position (ARSZ)	Momentum of Red on arrival at Blue position (MOM)	Average reduction in Red momentum per 1000 shells fired (ARMPKS)
Fire delivered on a distance criterion	light	4.31	9.87	1684.9	170.6	7.82
	heavy	4.43	10.31	1526.3	148.9	6.74
Fire delivered on a speed criterion	light	5.93	12.81	1303.2	101.7	10.68
	heavy	6.37	13.56	772.3	56.9	7.67
Fire delivered on a momentum criterion	light	6.81	15.18	1200.0	79.0	11.07
	heavy	6.62	13.56	787.5	58.0	8.07

- the momentum of the Red force on arrival at Blue's position, (this terminal momentum was calculated as the Red arrival size/Red arrival time)
- the average reduction in Red's momentum achieved per 1000 shells delivered by Blue.

Some overall conclusions can be drawn from this set of results. These will be stated here and explained in more detail in the next section.

Firstly, from Red's point of view, it would appear beneficial, in terms of improving its momentum of arrival at Blue's position, to always adopt the flexible, variable distance formation change point strategy in response to Blue's attacks.

A comparison of Tables 8.5a and 8.5b across all Blue strategies indicates that this Red strategy always resulted in the same or a shorter advance time and in a higher arrival momentum at the Blue position. Interestingly, the advantage of this variable distance strategy is slightly greater under light fire delivery than under heavy fire delivery by Blue.

In general, Red achieves these results by changing formation at the same time or later than under the fixed distance strategy. The exception to this result was for heavy fire delivery by Blue under a speed criteria.

Table 8.5(b) Results from armoured advance model: Red using a variable distance strategy for formation change

Red: variable distance strategy / Blue Strategy		Time to company column deployment (TTCCD)	Time for Red to reach Blue (ART)	Size of Red force arriving at Blue position (ARSZ)	Momentum of Red on arrival at Blue position (MOM)	Average reduction in Red momentum per 1000 shells fired (ARMPKS)
Fire delivered on a distance criterion	light	4.31	9.62	1685.4	175.1	6.43
	heavy	4.43	10.06	1529.9	152.0	6.40
Fire delivered on a speed criterion	light	6.18	12.31	1286.2	104.5	10.80
	heavy	6.25	13.56	790.0	58.3	7.59
Fire delivered on a momentum criterion	light	7.25	13.88	1189.0	85.6	11.60
	heavy	6.56	13.44	790.0	58.7	8.13

It was expected that the improvement in advance time under the variable distance formation change strategy, would be more than offset by a reduction in the size of the Red force arriving at the Blue position. However, a reduction only occurred in two of the six results shown when Blue deployed light fire on speed and momentum criteria. Even then the Red arrival momentum was still above that achieved under the fixed distance formation change strategy.

From Blue's point of view the objective is to reduce the Red arrival momentum. An examination of the arrival momentum figures in both Tables 8.5 (a and b) indicate that it is always preferable for Blue to deliver fire on a Red speed, rather than Red distance, criterion. However, the low proportion of the Red distance of advance over which fire is delivered by Blue is crucial to this conclusion. It would also appear that it was better still, under light fire delivery, for Blue to deliver fire on a criterion of Red momentum. However, this conclusion did not apply when Blue delivered heavy fire. In this case, delivery of fire on a speed criterion was marginally better.

In general, the benefit of heavy fire over light fire, although delivered at double the rate of light fire, was not as great as might have beeen expected.

The percentage reductions in Red arrival momentum were 13%, 44% and 31% respectively under the Blue distance, speed and momentum criteria for both the fixed and variable formation change strategies.

It can also be concluded from Table 8.5 that light fire, although not reducing Red's arrival momentum as much as heavy fire, is more effective per unit of ammunition delivered. This result can be seen in the column for average reduction of Red momentum per 1000 shells fired, in both Tables 8.5a and 8.5b.

THE UNDERLYING FEEDBACK MODEL

Whilst the foregoing model and results presentation may point to some conclusion directly, it is of interest to try to use the model to provide a deeper explanation of the results. It is possible to generate more general insight and understanding by developing a simplified, but very explicit, qualitative model of the feedback processes at work. Such a model will now be developed. However, a certain amount of abstraction is involved in the creation of such a diagram in this case, and the resultant model will be seen to lose its one to one correspondence with the physical reality modelled which existed in Figure 8.3.

Figure 8.7 shows an influence diagram of the underlying feedback structure of the model. It will be seen that five major feedback loops exist. The one around the left-hand side of the diagram (loop A) is a negative loop by which Red attempts to control (maintain) its speed. As actual speed declines (as a result of Blue fire) the planned distance to formation change is increased. This results in a later formation change, the maintenance of a high density of formation and a higher scheduled speed. Size erosion within a given formation will, however, take place. This is because as long as a high density formation is maintained, the productivity of Blue fire, and hence its effectiveness, remains high. As the size of the advance falls the planned distance to formation change is reduced. This effect can be traced out around the negative feedback loop on the right-hand side of the diagram (loop B).

The other three feedback loops in Figure 8.7 are all negative and relate, in turn, to the effect of Red's distance of advance on the productivity of Blue fire (loop C), the 'rate of recovery of speed' loop (loop D) by which Red's speed rises whenever Blue fire ceases and the Blue fire delivery (loop E), shown only for the speed criterion.

Figure 8.7 provides a clear explanation of the previously presented results and conclusions for Red strategy, which centres on the effect of the important, but rather inconspicuous loop in Figure 8.7, associated with speed recovery, and the way in which this relates to the strategy of the

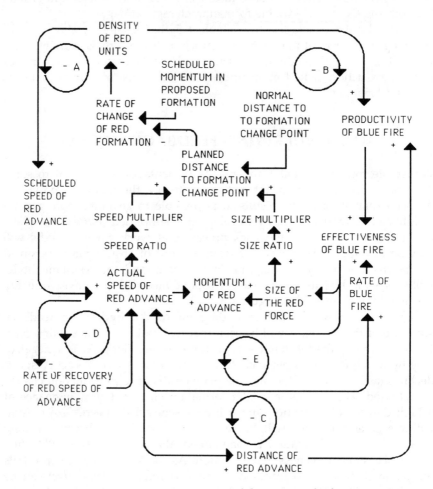

Figure 8.7 Underlying feedback structure of the armoured advance model

combatants. The basic insight generated is that speed and size as system variables have very different characteristics. The most important of these differences is that the former is recoverable by Red if Blue firing stops, but that the latter is not.

This characteristic is very clear from Figure 8.8 which shows the speed and numbers advancing plotted over time for the run involving the variable distance formation change strategy for Red and the strategy of heavy fire delivered on a momentum criterion for Blue.

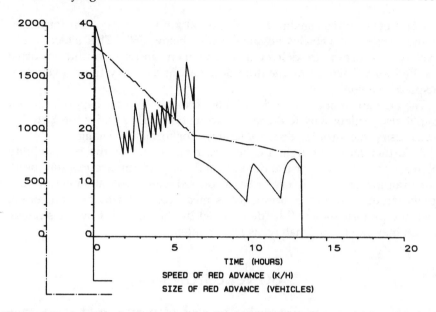

Figure 8.8 Armoured advance model. Variable distance formation change, high fire rate, momentum criterion

Consequently, momentum defined as the product of speed and size has a recoverable and non-recoverable component. Also, the level of the recovery is higher in a faster moving formation.

When Red delays its formation change it is preserving both a higher speed and a high level of ability to recover speed and, hence momentum lost during an attack. Thus, it is understandable that the variable distance formation change strategy is the strategy which gives the best results for Red in the model.

Understanding of the different characteristics of Red's size and speed is also vital to Blue. If Blue deploys intermittent fire delivery strategies of the type modelled, then this facilitates a recovery of Red speed and momentum. In particular, the strategies of delivering fire on a basis of Red's speed and momentum using heavy fire produces even more intermittent delivery patterns. This is because the levels of Red's speed and momentum at which fire is switched on and off, are reached much faster, hence, the advantage of heavy fire, over light, might be less than anticipated. The results suggest that continuous, rather than intermittent, fire delivery strategies are required. This issue is considered further in Chapter 10.

A further effect occurs when Blue delivers heavy fire on a momentum criterion. Here, high attrition takes place and this may be so great that Blue fire is switched off even when Red is travelling at its maximum speed. (The switch-on point is at a momentum value of 3000 (v^* (k/h)), which is not reached even at the maximum speed of 40 kmh^{-1} in battalion formation if the number of vehicles advancing falls below 750.) This effect would explain why heavy fire delivered on a momentum criterion did not drive the Red arrival size or momentum down as far as heavy fire delivered on a speed criterion.

The high attrition associated with Blue fire delivered on a speed criterion might also explain why Red changed formation earlier under this criterion when using the variable distance strategy for formation change.

A further factor brought to the fore here is that of the compatibility between the criterion for fire delivery and the performance measure used. For example, if fire delivery is based on reducing Red's momentum to a given target level and performance is measured in terms of momentum, then the performance will be determined by the target set by the strategy which may not be, ultimately, as low as intended.

CONCLUSIONS

This chapter has demonstrated a number of points about the System Dynamics methodology. Firstly, that in a large model it is vital to retain an abstracted picture of the underlying feedback structure of the model to aid analysis. Secondly, that models can be constructed to examine strategies between two competing factions. Thirdly, that a variety of performance measures can be utilised within a model to provide a summary of model behaviour and that, just like in the case of behaviour over time, changes in these can be explained in terms of the feedback structure of the model so as to enhance understanding.

Of particular interest was the identification of Red's speed as a more important parameter than Red's size, since it has the attribute of recoverability. Additionally important was the demonstration that intuitive strategies for Blue, such as delivering heavy fire in general, and heavy fire based on a criterion of Red's momentum in particular, could have much less than the anticipated effect.

REFERENCE

Huber, R.K. (1985) On Current Issues in Defence Systems Analysis and Combat Modelling, OMEGA International Journal of Management Science, 13 95–106.

Chapter 9

The Basic Concepts of System Dynamics Optimisation

INTRODUCTION

Traditionally, System Dynamics has relied on the use of intuition and experience by system owners and analysts to help design policies for improving system behaviour over time. Such an approach was described in Chapter 4 and extensively applied in the case studies, particularly in Chapter 7. This situation is now changing and much effort is being expounded in the development of more formal policy design methods.

Basically, two schools of thought are emerging. The first of these concerns the application of control theoretic methods such as control theory and optimal control theory (Sharma 1985, Mohapatra and Sharma 1985). These approaches can be powerful but do require substantial assumptions and a good level of analytical knowledge. Intensive use of these methods is not anticipated until sophisticated computer software is developed to improve their ease of application.

The second major approach to policy design which has emerged in recent years is that of simulation by optimisation (Keloharju 1981, Kivijavi and Tuominen 1986, Coyle 1985, Keloharju and Wolstenholme 1986, Wolstenholme and Al-Alusi 1987, Dangerfield and Roberts 1989). This approach also relies fundamentally on computer software, but is not inhibiting in its dependence on sophisticated analytical techniques.

Classical Quantitative System Dynamics as described so far in this book can be summarised in the following way. The process of model construction and development starts by considering a reference mode over time concerning the real world behaviour of the system of interest. From this, a feedback model is conceptualised to represent a dynamic hypothesis to explain this observed behaviour. A computer is then used as a fast calculating machine to check if the model can reproduce the reference mode behaviour and hence substantiate the hypothesis. Revisions to model parameters and structures are made manually until the model can achieve this objective. When it can, the model is considered validated and appropriate for use in designing other system behaviour modes by further parameter and structural changes. This process of model design is summarised as an iterative procedure in Figure 9.1.

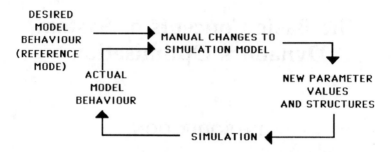

Figure 9.1   The process of model design in conventional System Dynamics

The concept of optimisation in System Dynamics is based on a belief that the manual procedure of system design given in Figure 9.1 can be automated by interfacing a heuristic optimisation algorithm with a System Dynamics simulation program. The combination of optimisation and simulation is not new in the modelling world. However, the approach is technically difficult and this factor has tended to inhibit its implementation. This chapter is concerned with describing the development and use of software aimed at making such a procedure available to a much wider range of analysts and system owners than has been previously possible. The optimisation algorithm used is the heuristic Search Decision Rule algorithm (SDR) (Buffa and Taubert 1972) and the System Dynamics software used is the original Dynamic System Modelling Analysis Programme (DYSMAP) (Cavana and Coyle 1982), the combined program being referred to as DYSMOD (Dynamic System Model Optimiser and Developer (Luostarinen 1982, Keloharju 1981)[1]. The process of model design using DYSMOD is outlined in Figure 9.2.

THE OPTIMISATION CONCEPT

Before applying the optimisation approach, two steps are necessary:
   Firstly, an objective function must be defined within the simulation model which summarises overall model behaviour. It is good practice to base each

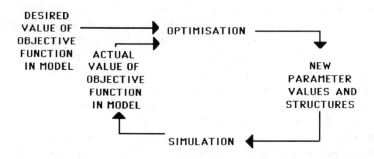

Figure 9.2 The process of model design in optimised System Dynamics

objective function on a level variable which is cumulated over a complete run. Such a function summarises a whole run of the model.

Secondly, a number of parameters within the model must be selected as candidates for optimisation, together with a range of feasible numerical values for each.

Each iteration of the procedure starts with a simulation run which calculates the value of the objective function chosen, under the initial conditions chosen for the simulation parameters. The SDR algorithm then treats these parameters as variables for optimisation and optimises them heuristically, that is, by changing them one at a time using the objective function as a measure of performance.

The output from the optimiser consists of a new set of parameter values. Subsequent iterations repeat this cycle beginning with the calculation during simulation of a new value of the objective function using the modified parameters. The algorithm compares the value of the objective function at the end of the simulation run to the corresponding value from the 'best' run completed to date. Although the objective function value may fluctuate from one iteration to another, the current optimal solution is always kept stored in the computer memory. This guarantees that optimisation never worsens the initial situation.

The modeller must decide on the number of iterations which are appropriate relative to achieving some desired value of the objective function. The number chosen is, however, not too important as the whole procedure can be subsequently continued, if necessary until the parameter values or objective function values stabilise. Typically 100 iterations might be required in a medium sized model.

Any existing parameters of the simulation model may be selected as candidate variables for optimisation. When a functional relationship is defined by a curve, it can be approximated with a table function in conventional System Dynamics. Table functions, as represented by piecewise linear curves, may also be chosen for optimisation. The program treats each separate corner-point of the curve as a separate parameter.

Great care must be exercised in both the selection of parameters and the choice of the feasible numerical range for each parameter to be optimised. Additionally, pseudo-parameters can be introduced into the simulation model for the purpose of optimisation. This is a powerful use of optimisation since it creates a means of carrying out structural rather than straight parameter optimisation, because the pseudo parameters can be employed as zero–one switching constants.

The following example shows how the process of optimisation occurs in parameter space. Consider a simple order backlog model as shown in Figure 9.3 and suppose that the production rate equation (shown in equation 9.1 without time subscripts) is the policy equation to be designed by optimisation. That is, the production rate is the objective function variable.

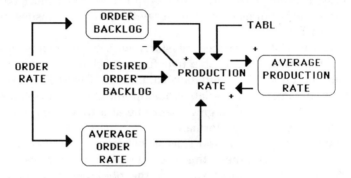

Figure 9.3 Influence diagram of a basic order backlog model

$$PR = (OBL-DOBL)/TABL+A1*AOR+(1-A1)*APR \qquad (9.1)$$

where

TABL=4,
A1=0.3,
AOR = average order,
APR = average production rate,
DOBL = desired order backlog,
OBL = order backlog,
PR = production rate and
TABL = time to adjust backlog.

Equation 9.1 contains two parameters, an existing model parameter (TABL) and an introduced or pseudo model parameter (A1) whose value could effectively change the structure of the model. These two parameters can be considered as candidate variables for optimisation to generate the 'best' form of the production rate equation.

The rectangle in Figure 9.4 defines the feasible parameter space for optimisation which is generated by the two parameters TABL and A1. It is obvious from Equation 9.1 that A1 must lie between 0 and 1. Suppose further that the modeller has defined the feasible range for TABL as being between 2 and 10.

The value of a chosen objective function guides the optimiser in search of more acceptable combinations of these two parameters. The path from the initial combination in Figure 9.4, however, depends on which objective function is chosen. Two hypothetical routes for model development are shown. The optimisation proceeds as follows. Some parameter combination is simulated, then changed by the optimiser within the given parameter ranges, and the new combination is simulated, etc. The procedure is called optimisation by simulation. Each parameter combination is a different model, therefore, the computer acts as a model generator.

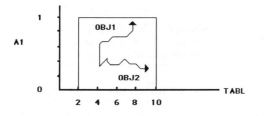

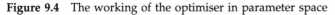

Figure 9.4 The working of the optimiser in parameter space

SOFTWARE FEATURES

The software development used in this book links the revised and extended Search Decision Rule (SDR) algorithm to the DYSMAP programming language (DYSMOD). Any System Dynamics model can thus be optimised without any changes to the format of the equations in the DYSMAP program. Figure 9.5 shows this software or frame as a black box. The flow chart in Figure 9.6 shows the most essential elements of the frame. The two boxes with a group of questions attached to them indicate the kind of information that the computer needs for optimisation. The answers to these questions are given to the program interactively.

The optimisation procedure begins with the nomination of the objective function. Optimisation is based on the value of the objective function at the end of the simulation run-length. Any model variable can be chosen as the

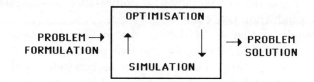

Figure 9.5 The frame as a black box

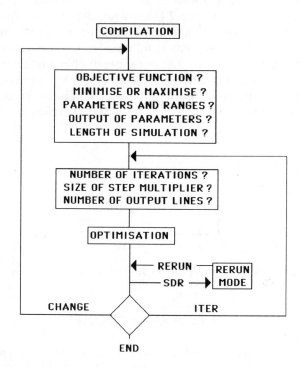

Figure 9.6 The simplified frame

objective function for optimisation and in practice the selection of a specific equation would have to be supported either by real life facts or some theory. The subsequent steps in the procedure are to give the program details of whether it is desired to maximise or minimise the objective function, to state the parameters which the optimiser may choose the values for, to give the upper and lower limits for each of these parameters and to decide on the number of iterations, the form of output and length of simulation.

As the optimisation algorithm changes parameters during the

optimisation process, statistical parameter data are generated. This information, which gives the parameter values as they stood at the end of each iteration, may be displayed as the optimisation process continues.

The way that the objective function value was developed during each iteration will always be printed out even though the user decides to disregard the parameter values. It is informative to see how the optimisation process is converging.

The SDR algorithm also utilises a technical parameter called size of step multiplier or, in short, step size. This parameter defines the amount of the initial change to be applied to each parameter value at the end of the first optimisation in terms of the maximum feasible change. Suppose that a parameter has an initial value of 10 and the range is from 2 to 20. If the step size is 0.3, the actual change at the end of the first optimisation would be $0.3*10=3$. Afterwards the algorithm changes the step size on the basis of the success achieved in the search process. That is, on the basis of the improvement achieved in the objective function between subsequent optimisations.

The box named 'optimisation' in Figure 9.6 refers to the optimisation process. When the process has ended, the computer prints out the final solution and then it lists the initial value and the best value of the objective function. Also the best values of the optimisation parameters are listed.

As shown in Figure 9.6, the modeller now has to choose from between four options.

(1) To let the computer generate some additional output data (such as graph plots) by rerunning the final optimised version of the model.
(2) To continue with the optimisation, that is to undertake more iterations.
(3) To change the objective function and/or the values of parameters and their ranges and continue optimising.
(4) To end the optimisation.

There are two modes available at this stage to achieve the above options; 'SDR' and 'Rerun'. The procedure is normally in the SDR-mode. In the 'Rerun' mode, required for (a) above, additional output data is available. To obtain it, the modeller has to type in the command RERUN followed by RUN. The computer then runs an additional simulation and prints out and/or plots the results. Simulation generates the results from the optimisation procedure because model parameters retain their values from optimisation.

However, if the optimisation process has not converged, as indicated by the presence of substantial changes in the objective function between consecutive optimisations, the optimisation process should be continued. By typing in SDR the computer returns to the SDR-mode. Then the option (b) can be chosen and further iterations carried out.

Alternatively, if the objective function did not work effectively during the optimisation process, then by typing in CHANGE the program transfers to the beginning of the algorithm and the objective function can be changed. Additionally, parameter values can be changed by typing in the corresponding parameter name.

OPTIMISATION IN ACTION

Having illustrated the method, two simple applications will now be presented. Example 1 shows how a curve fitting problem can be solved with DYSMOD using a mathematical objective function. Here the optimiser is used to carry out non-linear regression using an objective function which minimises the sum of squared deviations between the generated curve and given empirical data.

Example 2 develops a simple diffusion model. Here the purpose is to find a model which generates a given output. The objective function is the same as in example 1, but derives from the substance of the study.

Example 1: Curve Fitting

Suppose demand data from the last two years are as follows:

430/447/440/316/397/375/292/458/400/350/284/400/483/509/475/
500/600/700/700/725/600/432/615

In order to fit a curve to this data, the form of the demand equation has to be chosen. Suppose that the model consists of three additive components; a constant, a straight line and a sine wave. The first 'trick' of optimisation is to combine these in one demand equation as follows:

DEQ.K=CON+SLP*TIME.K-AMPL*SIN(6.283*TIME.K/PRD)   (9.2)

where

DEQ = demand equation,
CON = constant,
SLP = slope,
AMPL = amplitude of sine wave,
PRD = period of sine wave.

The demand equation has four parameters: CON, SLP, AMPL and PRD. Their values will emerge as a byproduct of the curve fitting procedure. Let the demand curve begin from the same point as the empirical data to be fitted to it. Therefore CON has the value of 430. AMPL has a minus-sign

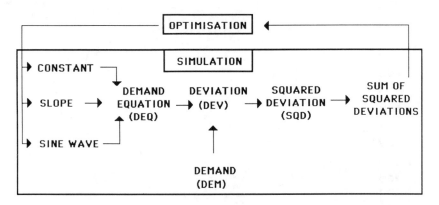

Figure 9.7   Influence diagram for curve fitting model

because empirical data indicated a cyclical decline during the first half-year of the study.

As seen from the influence diagram of Figure 9.7 the simulation model does not have any feedback loops. The simulation model is therefore open, but the optimiser itself creates a *feed-forward* path between the simulation output and input. The full equations for this model are given in Table 1 of Appendix 5.

All model listings in this chapter were created by transferring the DYSMOD models to DYSMAP2 (Dangerfield and Vapenikova 1987). Graph plots were then obtained using the PCPLOT software (Jackman 1988).

Example 1 has three purposes. It shows that:

(1) simulation in system dynamics is a special case of optimisation,
(2) the optimiser may find a good model even if the initial model is poor,
(3) the optimiser works as a model generator.

The model was first run without allowing any parameters to change. The optimiser cannot, of course, design the model unless parameters are allowed to change. For this reason such a run is referred to as dummy optimisation. In other words conventional simulation with fixed parameters can be considered as a special case of optimisation.

A plot of the base run of the dummy optimised model is presented in Figure 9.8 and shows clearly that the model is poor, therefore, it is necessary to allow the optimiser to choose the three parameters SLP, AMPL and PRD.

The ranges set for these parameters were (0.5–10), (50–300) and (10–200), respectively and the objective function to be optimised was the sum of squared deviations between the actual demand and value of the demand

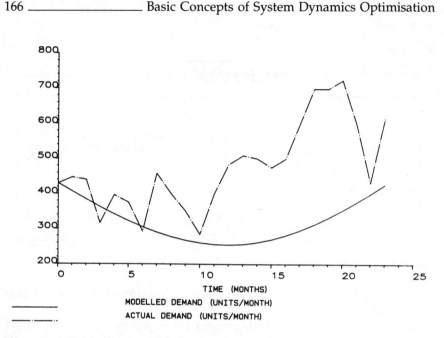

Figure 9.8 Curve fitting model, dummy optimisation

equation (SSD). The optimised values of the three parameters taken after 50 iterations were SLP = 3.998 AMPL=109.475 and PRD=26.060. The optimised values were used in the program listing in Table 1 of Appendix 5 to give the second run of the model and Figure 9.9 shows the graphical output obtained.

A comparison of the plottings from the initial model and from the final model shows that the demand equation DEQ has changed its shape and position. There were, however, many intermediate steps and any of them could have been plotted afterwards.

The use of optimisation in fitting model results to actual observations can be very powerful at the model validation stage and provokes comparisons between System Dynamics and Econometrics. However, care must be taken since, as explained in Chapter 4, a close fit between à model results and past actual data is only one aspect of validation in System Dynamics.

Nevertheless, such fitting of a model to real data can be a revealing process in determining the form of uncertain relationships in models.

One interesting application of the curve fitting concept was by Dangerfield and Roberts (1989) in their System Dynamics model of the spread of AIDS in the UK homosexual population. Predictions of the future incidence of AIDS depend crucially on numbers currently infected with HIV (Human Immuno-Deficiency Virus, the causative agent of AIDS) and who have not yet manifested those symptoms compatible with a

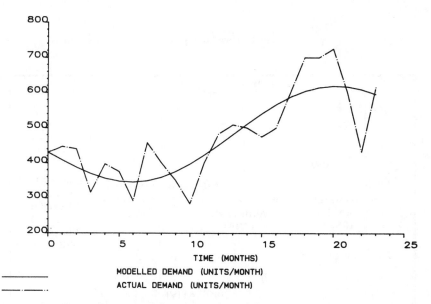

Figure 9.9 Curve fitting model, optimised model

confirmed diagnosis of AIDS. However, direct estimates of the prevalence
of HIV are extremely difficult to obtain because a large proportion of those
infected (seropositive) can be asymptomatic for many years. Dangerfield
and Roberts derived an optimised fit of the simulated to the observed
cumulative cases of AIDS and, once this was accomplished, they could
simply, inspect the values of the associated model variable defining the
prevalence of HIV.

Example 2: Behaviour Reproduction

A number of studies in the area of innovation adoption have shown that
a normal curve gives a good fit to the data which relates to the number
of adopters through time, that is, the cumulative number of new adopters
over time (CNA) follows an s-curve pattern.

Suppose that it is required to develop a model which produces output
corresponding to the s-shaped growth. Figure 9.10 shows an influence
diagram of how this might be constructed. An innovation diffusion rate
feeds a level of cumulative new adopters and it is required to generate
future diffusion rates from existing cumulative adopters. The question is
what shape should be attached to the intervening curve?

Table 2 in Appendix 5 gives a DYSMAP2 model listing which contains

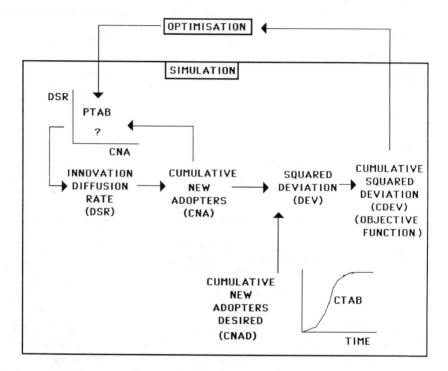

Figure 9.10 Influence diagram of behaviour reproduction model

the information of Figure 9.10. It will be noted that the relationship to be designed is specified as a constant table function and that simulation (dummy optimisation) of this model will result simply in a linearly increasing relationship between cumulative new adopters and time (Figure 9.11).

This model is poor as no use is made of the feedback theory. The s-formed output, in fact, requires the model to have a change in polarity of the feedback loop linking the diffusion rate and level of cumulative new adopters, therefore, a set of models, defined in CNA/DSR space, is required.

The objective function used is that of the cumulative squared deviations between the desired and actual cumulative new adopters. The parameters for optimisation are the five table function values defining the CNA/DSR relationship. The range of DSR values are defined as from 0 to 10 adopters per month. The optimiser can then be used to find out if the loop-polarity has to change before the model will be acceptable.

Such an optimisation experiment was carried out and the values chosen

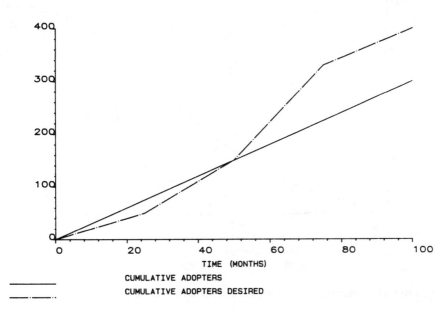

Figure 9.11 Innovation adoption model, dummy optimisation

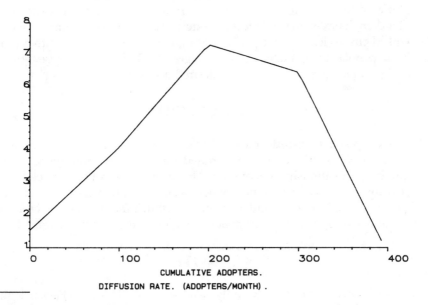

Figure 9.12 Innovation adoption model, optimised relationship between cumulative adopters and diffusion rate

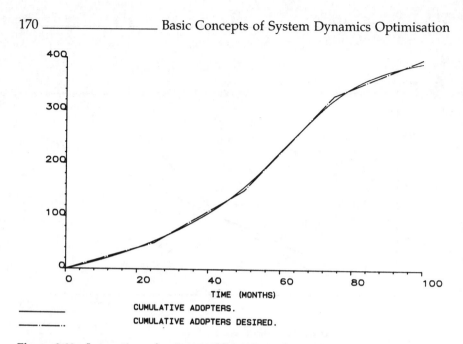

Figure 9.13   Innovation adoption model, optimised model

by the optimiser are shown in the table function of Figure 9.12 for the second run of the model in Table 2 in Appendix 5. Figure 9.12 is a scatter plot of diffusion rate (DSR) against cumulative new adopters (CNA). This plot demonstrates that when the model is run with the optimised values a kinked curve clearly emerges, indicating that a closed model with a change in loop polarity is indeed needed. Figure 9.13 shows the closeness of fit achieved between the actual and desired adopters.

CONCLUSIONS

This chapter has introduced the basic elements of heuristic optimisation in System Dynamics and demonstrated some simple applications in curve fitting and relationship generation. The approach can be described as one of automatic model generation, where a set of parameters can be designed and structured for a particular purpose. A much more extensive case study of the use of optimisation in policy design is presented in Chapter 10.

NOTE

[1] This software originally only available on Hewlett Packard mini-computers, has recently been implemented on 80386 based personal computers.

REFERENCES

Buffa, S. and W. H. Taubert (1972) *Production-Inventory Systems: Planning and Control* (Revised Edition), Irwin, Homewood, IL.

Cavana, R. Y. and R. G. Coyle (1982) *DYSMAP User Manual*, University of Bradford.

Coyle, R. G. (1985) The Use of Optimisation Methods for Policy Design in a System Dynamics Model, *System Dynamics Review*, **1**, 81–92.

Dangerfield B.C. and K. Roberts (1989) A Role for System Dynamics in Modelling the Spread of AIDS, *Transactions of the Institute of Measurement and Control*, **11**, 187–195.

Dangerfield, B. and O. Vapenikova (1987) *DYSMAP2 User Manual*, University of Salford.

Jackman, R. (1988) *PCPLOT User Manual*, University of Salford.

Keloharju R. and E. F. Wolstenholme (1989) A Case Study in System Dynamics Optimisation, *Journal of the Operational Research Society*, **40**, 101–110.

Keloharju, R. (1981) *Relativity Dynamics*. Helsinki School of Economics, Helsinki.

Kivijavi, H. and M. Tuominen (1986) Solving Economic Optimal Control Problems with System Dynamics. *System Dynamics Review*, **2**, 138–150

Luostarinen, A. (1982) *DYSMOD User Manual*, University of Bradford.

Mohapatra, P. K. J. and S. D. Sharma (1985) Synthetic design of policy decisions in system dynamics models: a modal control theoretical approach, *System Dynamics Review*, **1**, 63–81.

Sharma, S. K. (1985) *Policy Design in System Dynamics Models: Some Control Theory Applications*, Doctoral Thesis submitted to the Indian Institute of Technology, Kharagpur, India.

Wolstenholme, E. F. and A-S. Al-Alusi (1987) System Dynamics and Heuristic Optimisation in Defence Analysis, *System Dynamics Review*, **3**, 102–116.

Chapter 10

The Analysis of Defence Strategies Using Optimisation

INTRODUCTION

The aim of this chapter is to demonstrate the merits of using optimisation for policy design in a System Dynamics model by applying the approach to the Armoured Advance Model developed in Chapter 8 and listed in Appendix 4. It is assumed here that the reader has read the background to the construction and analysis of this model.

The rationale for the choice of objective functions and optimisation parameters are given. Particular attention is paid to the contribution of the optimisation process in highlighting new interactions between Red and Blue strategies and in adding to the previous insights generated by conventional analysis.

THE OPTIMISATION OF RED STRATEGIES

The approach in applying optimisation to the armoured advance model has been to take each previously used performance measure in turn as an objective function. Then, to define for each objective function the parameters whose values will be chosen by the optimisation procedure, and, additionally, the upper and lower limits of the feasible ranges for these values.

In general the guideline followed in defining parameters was that only Red's strategy parameters could be involved in experiments using Red's objective functions and only Blue's strategy parameters could be involved in experiments using Blue's objective functions.

A detailed definition of each of Red's objective functions and the corresponding definition of parameters and their range settings used in each optimisation experiment with the model is given in Table 10.1.

From Red's point of view the main objectives are to maximise its size and momentum on arrival at the Blue position or to minimise its total advance time (experiments 1 to 3 in Table 10.1). For all three of these experiments, the relevant optimisation parameters are those involving the Red formation change decisions.

Table 10.1. Definition of experimental design for Red's objective functions

Experiment No.	Red's objective function	Parameters (and their ranges) involved in the optimisation
1.	Maximise Red arrival size	Speed/size switch (0–1). The y coordinates of the *size* multiplier table (Figure 8.5) (0–1)
2.	Minimise Red arrival time	Speed/size switch (0–1). The y coordinates of the *speed* multiplier table (Figure 8.4) (1–2)
3.	Maximise Red arrival momentum	Speed/size switch (0–1). The y coordinates of the *size* and *speed* multiplier tables (0–1) (1–2)

The first of these is the Red speed–size switch defined in equation 8.1. It should now be appreciated that allowing the optimisation procedure to choose the value of this parameter, between 0 and 1 is equivalent to choosing whether Red speed or size should be used as a basis for modifying the distance to company column deployment.

The second set of parameters are the y coordinates of the Red speed and size multiplier tables, also defined in equation 8.1 and in Figures 8.4 and 8.5 in Chapter 8. These tables determine how the Red distance to company column deployment will vary with deviations in speed and size from those scheduled. Allowing the optimisation procedure to choose the values of these coordinates, within the ranges defined in Table 10.1, is equivalent to choosing the shape of the multiplier functions.

For the objective function of minimising Red arrival time only the y coordinates of the *speed* graph were allowed to be chosen. For the objective function of maximising Red arrival size only the y coordinates of the *size* graph were allowed to be chosen. For the objective function of maximising Red arrival momentum the y coordinates of *both* graphs were allowed to be chosen.

RESULTS FROM EXPERIMENTS WITH RED'S OBJECTIVE FUNCTIONS

Note that, since the Red objective function experiments involve choosing the Red distance to company column formation, they only involve the Red *variable* distance strategy and there are no results corresponding to the previous *fixed* distance strategy.

Maximising Red's Arrival Size

Table 10.2 presents a set of results (final optimised values) from maximising Red's arrival size. A full comparison of these results with the original results can be achieved by comparing Table 10.2 with Table 8.5b. For convenience, however, the original value of the arrival size for each of Blue's fire delivery criterion is given in Table 10.2.

In every case, the optimised arrival size significantly exceeds the original arrival size. This is achieved by the optimiser choosing Red *size* as the sole determinant of the formation change point (see SZS column in Table 10.2) and adjusting the shape of the size multiplier to produce a much more aggressive response in the planned distance to the formation change point, as soon as the actual Red size falls below the planned Red size.

The consequence is that, in the cases of Blue's fire delivered on speed and momentum criteria, the whole of the Red advance (apart from the first DT of the simulation) is carried out in company columns. In the case of Blue's fire delivery on a distance criterion, the time spent by Red in battalion formation is slightly longer, since, by definition of the strategy, Blue's fire is much less in total. In all cases, the focus on a long advance in company columns means that Red's total advance times are much slower and Red's arrival momentum is much worse than in the original model. Additionally, Blue uses much more ammunition and the efficiency of its use is correspondingly low.

This result, of carrying out the whole advance in company columns, represents the maximum extent to which Red can protect itself against losses incurred in force size. It might be expected that the gain in the size of the Red force on arrival at the Blue position would at least compensate for the longer advance time and, hence, not seriously affect the momentum of Red's arrival at Blue's position. However, as revealed by the actual arrival momentum figures in Table 10.2, this is not the case and there is a significant deterioration in the values for Red momentum on arrival at the Blue position relative to the results for this variable in Table 8.5b.

As in Chapter 8, an explanation of the result is that an early formation change by Red means a lower average speed of advance and a reduced rate of speed recovery when Blue firing ceases. However, the extreme situation

Table 10.2 Results of maximising Red's arrival size

Blue	Red:	Time to company column deployment (TTCCD)	Time for Red to reach Blue (ART)	Size of Red force arriving at Blue position (ARSZ)	Momentum of Red on arrival at Blue position (MOM)	Average reduction in Red momentum per 1000 shells fired (ARMPKS)	Size of Red force arriving at Blue position (from Table 8.5b) (ART)	(SZS)
Fire delivered on a distance criterion	light	3.37	11.06	1710.0	154.58	11.51	1685.4	0
	heavy	3.50	11.62	1584.2	136.27	7.10	1529.9	0
Fire delivered on a speed criterion	light	0.12	21.31	1624.8	76.23	8.01	1286.2	0
	heavy	0.12	22.50	1421.0	63.64	4.09	790.0	0
Fire delivered on a momentum criterion	light	0.12	26.18	1545.9	59.03	7.40	1189.0	0
	heavy	0.12	31.56	1245.7	39.46	3.39	790.0	0

Table 10.3 Results of minimising Red's arrival time

Blue	Red:	Time to company column deployment (TTCCD)	Time for Red to reach Blue (ART)	Size of Red force arriving at Blue position (ARSZ)	Momentim of Red on arrival at Blue position (MOM)	Average reduction in Red momentum per 1000 shells fired (ARMPKS)	Time for Red to reach Blue (From Table 8.5b) (ART)	(SZS)
Fire delivered on a distance criterion	light	6.18	6.18	1721.2	278.1	9.81	9.62	1
	heavy	6.37	6.31	1612.2	255.4	8.52	10.06	1
Fire delivered on a speed criterion	light	7.62	10.56	1098.5	103.9	12.77	12.31	1
	heavy	6.87	13.10	911.1	59.9	6.47	13.56	1
Fire delivered on a momentum criterion	light	9.87	9.87	879.9	89.1	17.06	13.88	1
	heavy	8.71	8.75	707.3	60.8	12.06	13.44	1

chosen by the optimisation process, highlights an additional contributory factor. A total advance in company columns not only results in a slow speed which increases the time of the advance, but it also results in an increase in the period over which Red is exposed to Blue fire.

Minimising Red's Arrival Time

Table 10.3 presents the set of results (final optimised values) from minimising Red's arrival time. A full comparison of these results, with the original results, can be achieved by comparing Table 10.3 with Table 8.5b. Again, for convenience, the original value of the arrival time for each of Blue's fire delivery criterion is given in Table 10.3.

It will be seen in every case that the optimised arrival time is significantly less than that recorded from the original model. This improvement is achieved in the optimisation process by employing *speed* as the sole determinant of the formation change point (see SZS column in Table 10.3) and by again adjusting the shape of the speed multiplier. Once again, there is a much more aggressive response in the planned distance to the formation change point, as soon as the actual Red's speed falls below the planned speed. The consequence is that almost the whole of the Red advance is now carried out in battalion formation.

Surprisingly, the Red arrival size does not suffer too greatly and, with only one exception, a higher arrival momentum is achieved by Red than in the original results of Table 8.5b. Blue's use of ammunition is also low and its efficiency of use high. It would appear, therefore, that the gain in speed achieved by Red by staying in battalion formation more than compensates for the higher attrition. This result can again be explained in terms of Red's increased ability to achieve a higher speed and rate of speed recovery by maintaining a battalion formation. However, the extreme case of Red choosing a total advance in battalion formation again highlights, additionally, the advantage that this might have in reducing the Red exposure time to Blue fire.

Maximising Red's Arrival Momentum

Table 10.4 presents the results (final optimised values) from maximising Red's arrival momentum at the Blue position. A full comparison of these results with the original results can be achieved by comparing Table 10.4 with Table 8.5b. Again, for convenience, the original value of the Red's momentum for each of Blue's fire delivery criterion is given in Table 10.4.

In every case the optimised value of Red's arrival momentum is higher than that recorded in the results from the original model. The results tend strongly towards those seen in the experiments to minimise Red arrival

Table 10.4 Results of maximising Red's arrival momentum

Blue	Red:	Time to company column deployment (TTCCD)	Time for Red to reach Blue (ART)	Size of Red force arriving at Blue position (ARSZ)	Momentum of Red on arrival at Blue position (MOM)	Average reduction in Red momentum per 1000 shells fired (ARMPKS)	Momentum of Red on arrival at Blue position (From Table 8.5b) (MOM)	(SZS)
Fire delivered on a distance criterion	light	6.18	6.18	1721.2	278.1	9.81	175.1	1.0
	heavy	6.37	6.31	1612.2	255.4	8.52	152.0	1.0
Fire delivered on a speed criterion	light	6.18	12.06	1284.2	106.4	10.78	104.5	0.8
	heavy	4.87	15.18	911.1	59.9	6.47	58.3	1.0
Fire delivered on a momentum criterion	light	8.18	12.12	1093.6	90.1	12.75	85.6	0.9
	heavy	8.81	8.75	707.3	80.8	12.06	58.7	1.0

time, rather than towards those seen in the experiment to maximise Red's arrival size. The optimiser predominantly chooses Red *speed* as the criterion for formation change with the shape of both the speed and *size* multiplier graphs chosen for aggressive responses. There is a clear emphasis on *delaying* formation change by Red, but not as much as encountered in the experiments involving the objective function of minimising Red's arrival time. Advance times are, however, much shorter than in the original model and Red's arrival size is not badly affected.

CONCLUSIONS FROM OPTIMISING RED'S STRATEGIES

These results provide substantial confirmation of the previous conclusions from the non-optimised model, that Red's best strategy is to stay in battalion formation for as long as possible.

The investigation of what might be intuitively considered as trivial runs from two extreme situations (maximising Red's arrival size and minimising Red's arrival time), has been shown to facilitate the formation of a perspective concerning the trade off between the two.

This perspective is confirmed in the run involving maximisation of Red's arrival momentum. The results from this run can be clearly seen to tend towards the results from minimising Red's arrival time rather than maximising arrival size.

The approach has added to the analysis by clearly highlighting a further effect by which Blue's strategies interact with Red's. Originally the Red desire to stay in battalion formation was seen primarily as being based on increasing its ability to recover speed, and hence momentum, when Blue's firing ceased. Additionally, however, as shown here, it is clear that this strategy further minimises the time over which Red is exposed to Blue's fire.

These two effects are demonstrated in the influence diagram of Figure 10.1. Figure 10.1a shows the two (speed and size) negative control loops by which Blue controls Red's speed and size by adjusting the rate of fire. Figure 10.1b shows how certain fire strategies (like switching fire off when momentum reaches a certain point) can trigger the recovery of Red's speed. Figure 10.1c shows how a positive speed loop can be created by taking into account Red's exposure times to Blue's fire. When Red maintains speed by holding its battalion formation the total advance time and, hence, exposure time to Blue's fire is reduced. Blue's cumulative fire delivered is less and hence Red's speed, size and momentum are higher.

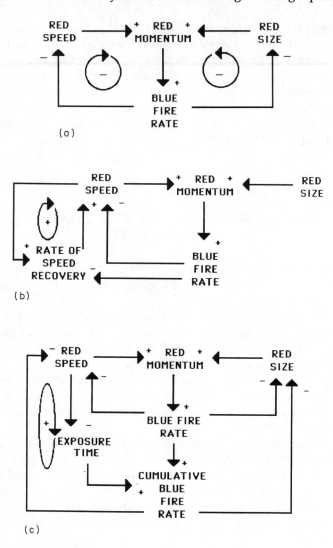

Figure 10.1 (a) The normal control of Red size and speed by Blue fire, (b) Red speed recovery brought about by a Blue strategy of switching off fire, (c) Red speed and size preservation achieved by reducing the time of Red exposure to Blue fire

THE OPTIMISATION OF BLUE STRATEGIES

From Blue's point of view its main objectives are the opposite to those defined for Red. These are, to minimise Red's arrival size, maximise Red's arrival time and to minimise Red's arrival momentum (experiments 1, 2 and 3 in Table 10.5). Additionally, it has the aim of trying to achieve these objectives, particularly that of minimising Red's momentum with the minimum use of ammunition. Experiment 4 is, therefore, constructed for this purpose. This involves an objective function of maximising the average reduction in Red momentum on arrival at the Blue position per 1000 shells fired.

For all experiments, the relevant parameters to be chosen are those involving Blue fire delivery: (i) the proportion of the Red advance over which Blue delivers fire (where the latter is carried out on a distance criteria); (ii) the upper and lower limits of Red speed at which Blue fire is switched on and off (where Blue fire is delivered on a speed criterion); and (iii) the upper and lower limits of Red momentum at which Blue fire is switched on and off (where Blue fire is delivered on a momentum criterion). The specific figures set for these ranges are given in Table 10.5.

Table 10.5 Definition of experimental design for blue objective functions

Experiment number	Blue Objective Functions	Parameters (and their ranges) involved in the optimisation
		All Blue fire delivery parameters:
1	Minimise Red Arrival Size	*On distance criterion*—the proportion of the distance of advance over which firing
2	Maximise Red Arrival Time	can take place (0–100%). *On speed criterion*—the upper and lower limits of Red speed (kmh^1) in battalion
3	Minimise Red Arrival Momentum	formation at which Blue firing was switched on (26–40) and off (10–25) and equivalent limits in company columns (10.5–14 and 5–9.5).
4	Maximise the Average Reduction in Red Momentum per 1000 Blue Shells fired	*On momentum criterion*—the upper and lower limits of Red Momentum in battalion formation at which Blue firing was switched on (25 500–45 000) and off (10 000–24 500) and equivalent limits in company columns (9500–16 000 and 2000 to 8500)

RESULTS FROM EXPERIMENTS WITH BLUE'S OBJECTIVE FUNCTIONS

The experiments listed in Table 10.5 were carried out in respect of both Red's fixed and variable formation change strategies. The results of these will be discussed separately.

Red Fixed Distance Formation Change Strategy

Table 10.6 presents a set of results (final optimised values), which are the same for the first three of Blue's objective functions (experiments 1, 2 and 3). It would appear that if Red's flexibility concerning its choice of formation change point is removed, then the same effect is achievable by Blue via any of these objectives.

Table 10.6 also contains the results obtained in Table 8.5a for comparison. It will be seen that in the optimised results Blue increases the time for Red to reach Blue and reduces the size of the Red force arriving at the Blue position and the Red arrival momentum in all cases.

Under the Blue strategy of delivering fire on a distance criterion, optimisation of Red's arrival size, time and momentum is achieved by firing during the whole of the Red advance, that is, by deploying continuous fire. Under the Blue strategy of fire delivery on a speed or momentum criteria, optimisation of Red arrival size, time and momentum is achieved by switching fire on and off at the lowest points defined for this on the range of speed or momentum defined.

Overall Blue achieves the longest time for Red's advance and the lowest values of Red's arrival speed and momentum when delivering fire on a distance criteria. This is because such a strategy allows continuity of fire delivery. As shown in Chapter 8, when fire is delivered on a speed or momentum criteria, there are periods when fire is switched off, which facilitates Red speed recovery. It will be noted from Table 10.6 that under a heavy fire delivery rate and a Blue distance criterion for fire delivery that Red is totally attrited whilst still in battalion formation.

Table 10.7 presents a set of results from experiment 4, which corresponds to the maximisation of the reduction in Red's arrival momentum per 1000 shells fired. The improvement in this parameter over the original model is achieved by Blue only choosing to fire whilst Red is in battalion formation.

This result constitutes an interesting insight and arises because the density of this formation is higher than a company column formation and, hence, the productivity of the Blue fire is increased. However, Red's arrival size, time and momentum are all improved relative to experiments 1 to 3 because less fire is delivered overall, which is, of course, detrimental to Blue. Nevertheless, the savings in ammunition can be phenomenal, which might more than compensate for the increase in Red's arrival momentum if ammunition is limited.

As observed in Chapter 8, the greatest reduction in momentum per 1000 shells fired, is achieved by Blue deploying light rather than heavy fire. This is because not only does light fire save on ammunition, but also it results in more consistent fire. The effect is particularly noticeable when

Table 10.6 Results from maximising Red arrival time and minimising Red arrival size and momentum under a fixed distance formation change strategy

Blue	Red:	Time to company column deployment (TTCCD)	Time for Red to reach Blue (ART)	Size of Red force arriving at Blue position (ARSZ)	Momentum of Red on arrival position (MOM)	Average reduction in Red momentum per 1000 shells fired (ARMPKS)	Size of for arriving at Blue position (ARSZ)	Time for Red to reach Blue position (From Table 8.5a) (ART)	Momentum of Red on arrival at Blue position (MOM)
Fire delivered on a distance criterion	light	14.18	22.56	566.6	25.11	10.68	1684.9	9.87	170.6
	heavy	∞	∞	0	0	10.76	1526.3	10.31	148.9
Fire delivered on a speed criterion	light	6.50	14.06	1234.6	87.79	11.06	1303.2	12.81	101.7
	heavy	7.43	16.18	569.2	35.16	7.31	772.3	13.56	56.9
Fire delivered on a momentum criterion	light	7.81	16.40	1123.2	68.07	11.27	1200.0	15.18	79.0
	heavy	8.06	15.56	553.2	35.55	8.02	787.5	13.56	58.0

Table 10.7 Results from maximising the average reduction in Red momentum per 1000 shells fired under a fixed distance formation change strategy

Blue	Red:	Time to company column deployment (TTCCD)	Time for Red to reach Blue position (ART)	Size of Red force arriving at Blue position (ARSZ)	Momentum of Red on arrival at Blue position (MOM)	Average reduction in Red momentum per 1000 shells fired (ARMPKS)	Average reduction in Red momentum per 1000 (From Table 8.5a) (ARMPKS)
Fire delivered on a distance criterion	light	8.12	13.50	1233.5	91.37	19.49	7.82
	heavy	6.06	11.43	1346.5	117.73	18.63	6.74
Fire delivered on a speed criterion	light	6.50	13.18	1254.9	95.15	11.22	10.68
	heavy	7.43	14.00	681.8	48.69	8.44	7.67
Fire delivered on a momentum criterion	light	7.81	14.62	1164.3	79.61	11.76	11.07
	heavy	7.68	13.06	843.4	64.57	9.10	8.07

Blue deploys fire on speed and momentum criteria when, as explained in Chapter 8, heavy fire is most intermittent.

Red Variable Distance Formation Change Strategy

The basic rule concerning the choice of parameters for optimisation of Blue's objective functions was similar to that quoted for Red objective functions, that is, only parameters under Blue's control should be used in experiments involving Blue's objective functions.

Given the significance of the Red variable distance formation change strategy, it was considered of interest to repeat the four experiments defined in Table 10.5 for this strategy. It is important to realise that this situation corresponds to Red being *manipulated* by Blue (since the outcome is controlled by a Blue objective function), rather than Red *adapting* to Blue's actions.

In the case of experiment 1 (minimising Red's arrival size) the results (Table 10.8) show that the manipulation of Red by Blue takes the form of Red changing formation late. This suggests that Blue would want to manipulate Red into spending as long a time as possible in the denser battalion formation to increase the productivity of its fire.

In the case of experiment 2 (maximising Red's arrival time) the results (Table 10.9) conversely show that Blue would want Red to change formation to the slower company column formation as early as possible, so as to prolong the total advance time. This is a similar result to that obtained when maximising Red's arrival size under the Red objective function experiments and is detrimental to Red.

In the case of experiment 3 (minimising Red's arrival momentum) the results are almost exactly the same as those of experiment 2, that is, Red changes formation early. This result suggests that manipulation of Red's speed is much more effective than manipulating its size and leads to a much more effective reduction in its arrival momentum. In fact manipulation of Red's arrival size by Blue plays into Red's hands since it promotes a lengthy advance in battalion formation which, as shown in the experiments on Red's objective functions, is Red's most effective strategy from its own point of view.

Overall Red's desire is to extend its battalion formation advance (where higher speed and less exposure time outweigh the higher attrition), whereas Blue would wish to extend its company column advance (where lower speed and longer exposure time outweigh the lower attrition).

Table 10.10 presents the results from maximising Red's reduction in momentum per 1000 shells fired. The predominant effect here is for Blue to manipulate Red into a late change of formation. Whilst this might be seen as a contradiction to the foregoing logic, it is entirely sensible under this

Table 10.8 Results from minimising Red's arrival size at the Blue position with Red's formation change strategy manipulated by Blue

Blue	Red:	Time to company column deployment (TTCCD)	Time for Red to reach Blue position (ART)	Size of Red force arriving at Blue position (ARSZ)	Momentum of Red on arrival at Blue shells fired (MOM)	Average reduction in Red momentum per 1000 shells fired (ARMPKS)	Size of Red force arriving at Blue position (from Table 8.5b) (ARSZ)
Fire delivered on a distance criterion	light	5.37	14.43	1287.5	89.17	10.44	1685.4
	heavy	6.06	30.00	884.6	29.48	4.29	1529.9
Fire delivered on a speed criterion	light	5.87	13.81	1277.2	92.46	10.72	1286.2
	heavy	5.43	15.56	796.6	51.19	6.68	790.0
Fire delivered on a momentum criterion	light	6.75	15.93	1194.5	74.94	10.95	1189.0
	heavy	6.00	17.31	711.2	41.08	6.48	790.0

Table 10.9 Results from maximising Red's arrival time and minimising Red's arrival momentum at the Blue position with Red's formation change strategy manipulated by Blue

Blue	Red:	Time to company column deployment (TTCCD)	Time for Red to reach Blue (ART)	Size of Red force arriving at Blue position (ARSZ)	Momentum of Red on arrival at Blue position (MOM)	Average reduction in Red momentum per 1000 shells fired (ARMPKS)	Time for Red force to reach Blue (ART) (From Table 8.5b)	Momentum of Red on arrival at Blue position (MOM)
Fire delivered on a distance criterion	light	3.87	18.50	1341.0	72.48	9.49	9.62	175.1
	heavy	2.06	30.00	892.9	29.76	3.65	10.06	152.0
Fire delivered on a speed criterion	light	3.87	16.68	1376.7	82.50	9.55	12.31	104.5
	heavy	2.06	23.00	1119.9	68.01	4.39	13.56	58.3
Fire delivered on a momentum criterion	light	3.87	19.34	1325.2	68.18	9.28	13.88	85.6
	heavy	2.06	30.00	884.6	29.48	3.63	13.44	58.7

Table 10.10 Results from maximising the average reduction in Red's arrival momentum per 1000 shells fired with Red's formation change strategy manipulated by Blue

Blue	Red:	Time to company column deployment (TTCCD)	Time for Red to reach Blue (ART)	Size of Red force arriving at Blue position (ARSZ)	Momentum of Red on arrival at Blue position (MOM)	Average reduction in Red momentum per 1000 shells fired (ARMPKS)	Average reduction in Red momentum per 1000 shells fired (From Table 8.5b) (ARMPKS)
Fire delivered on a distance criterion	light	3.87	12.93	1532.6	118.46	27.87	6.43
	heavy	4.20	14.37	1536.1	106.86	28.66	6.40
Fire delivered on a speed criterion	light	7.00	11.87	1211.7	102.40	11.99	10.80
	heavy	6.25	11.12	803.4	72.21	8.75	7.59
Fire delivered on a momentum criterion	light	8.12	12.75	1115.9	87.52	12.80	11.60
	heavy	7.00	11.37	879.3	77.30	13.39	8.13

objective function. It follows from the desire by Blue to keep Red in the high density battalion formation for as long as possible to improve its utilisation of ammunition. When company column formation is achieved by Red, no further Blue fire takes place. Again there is evidence that light fire by Blue is more effective than heavy.

In addition the optimisation procedure led to a further intriguing insight under experiment 4. Blue chose only to fire on Red at the most productive point during the Red advance. This was chosen by the optimiser to be at the very end of Red's battalion formation advance and this choice is explainable since it is when Red is not only in its most dense formation, but also at its closest point to Blue (hence Blue's accuracy of fire is higher). This result provides a good example of how optimisation facilitates an holistic appreciation and interpretation of results.

CONCLUSIONS FROM OPTIMISING BLUE'S STRATEGIES

It is concluded that extending the optimisation analysis of the armoured advance model to include Blue's objective functions provides additional insights as follows:

- It is important for Blue to maintain continuity of fire. This is facilitated by Blue firing on a distance rather than a speed or momentum criteria, since the latter results in periods of zero fire when the trigger points, defined in these fire delivery strategies, come into play. This in turn allows Red to recover speed which shortens the advance time.
- When ammunition is limited it is more productive for Blue to restrict fire to periods when Red is advancing in battalion formation, to apply light but consistent fire and to attack Red towards the end of its advance in battalion formation.
- The results from the situation where Blue is allowed to manipulate Red's formation change point, confirm the previous result from the experiments on Red's objective functions, that is, Red's speed is a more important variable than Red's size to both sides. It is best for Red always to extend its battalion formation advance as long as possible. This increases speed and, by reducing the exposure time to Blue fire, reduces attrition.
- When ammunition is limited it is to Blue's advantage if Red does prolong its advance in battalion formation, since Blue can fire for longer in its more productive mode.

OVERALL CONCLUSIONS

This chapter has shown that optimisation can add further value to the insights achieved from conventional model analysis. As in the step from Qualitative to Quantitative System Dynamics, there is a learning overhead involved in moving from Quantitative System Dynamics to Optimisation analysis. This is centred on the acquisition of skills in using additional computer software and the return obviously depends again on both the type of system under study and on the ability of the analyst.

The conclusions from the conventional System Dynamics analysis of the Armoured Advanced Model, given in Chapter 8, were based on 12 basic simulation runs. Here, the analysis was based on 100 iterations of 12 runs for each of seven objective functions, that is, 8400 runs. However, most of these were automatically carried out by the software. To achieve the same depth of analysis by conventional means would have required a repeat of the 12 original model runs for all of the permutations of the parameter values given in Tables 10.1 and 10.5, hence, the application of the optimiser subsumes a great deal of traditional sensitivity analysis.

In particular, the optimisation procedure will often seek out extreme values of parameters to achieve improved system behaviour. This was clearly demonstrated in its choice of Red strategies to maximise Red arrival size or to minimise Red arrival time at the Blue position by carrying out the whole of the advance in company columns or battalion formation, respectively.

In addition, to its role in policy analysis, the thoroughness of the approach is very useful at the model validation stage of the System Dynamics methodology and is equivalent to the idea of testing models to extremes. In newly created models the application of the procedure will often result in the optimisation of an objective function via routes which should not be allowed to exist in the model. These are, hence, identified and can be eliminated.

Appendix 1

DYSMAP2 Equations for the Models used in Chapter 5

(X or / * denotes a continuation line)

Table 1 Model SRL1

```
NOTE
NOTE    DYSMAP2 EQUATIONS FOR RUNS 1 TO 5 OF THE
NOTE    STAFF RECRUITMENT AND LEAVING (SRL) MODEL
NOTE        OF CHAPTER 5
NOTE        (RESULTS IN FIGURES 5.2 - 5.5)
NOTE
NOTE
NOTE  DIMENSIONLESS VARIABLES ARE GIVEN A DIMENSION OF
NOTE  (1)
NOTE
L STAFF.K=STAFF.J+DT*(RECRUIT_RATE.JK-LEAVING_RATE.JK)
N STAFF=INITIAL_STAFF
R LEAVING_RATE.KL=STAFF.K*PROPORTION_LEAVING
C PROPORTION_LEAVING=0.05
C INITIAL_STAFF=0
R RECRUIT_RATE.KL=(TARGET_STAFF-STAFF.K)/RECRUIT_DELAY
C RECRUIT_DELAY=5
C TARGET_STAFF=30
SPEC LENGTH=25/DT=0.1/PRTPER=1
PRINT 1)STAFF
PRINT 2)TARGET_STAFF
N TIME=0
NOTE
```

```
NOTE   DOCUMENTATION OF VARIABLES
NOTE
D STAFF=(PEOPLE)    NUMBER OF CURRENT STAFF MEMBERS
D RECRUIT_DELAY=(MONTH)    AVERAGE TIME TO RECRUIT STAFF
D RECRUIT_RATE=(PEOPLE/MONTH)    RATE OF RECRUITMENT OF
/* STAFF
D TARGET_STAFF=(PEOPLE)    TARGET NUMBER OF STAFF
D TIME=(MONTH)    SIMULATION TIME
D LENGTH=(MONTH)    SIMULATION LENGTH
D PRTPER=(MONTH)    SIMULATION PRINT PERIOD
D DT=(MONTH)    SIMULATION INTERVAL
D INITIAL_STAFF=(PEOPLE)    INITIAL NUMBER OF STAFF
D LEAVING_RATE=(PEOPLE/MONTH)    RATE OF LEAVING OF STAFF
D PROPORTION_LEAVING=(1/MONTH)    PROPORTION LEAVING PER
/*   MONTH
NOTE
NOTE ***********************************
RUN 1 SRL MODEL WITH PROPORTIONAL CONTROL
NOTE ***********************************
NOTE
C  RECRUIT_DELAY=2
NOTE
NOTE ******************************************
RUN 2 SRL MODEL WITH RECRUIT DELAY = 2 MONTHS
NOTE ******************************************
NOTE
C  RECRUIT_DELAY=10
NOTE
NOTE ********************************************
RUN 3 SRL MODEL WITH RECRUIT_DELAY = 10 MONTHS
NOTE ********************************************
NOTE
C RECRUIT_DELAY=5
L A_LEAVING_RATE.K=A_LEAVING_RATE.J+(DT/SMT)*
/*   (LEAVING_RATE.JK-A_LEAVING_RATE.J)
N A_LEAVING_RATE=LEAVING_RATE
C SMT=3
R RECRUIT_RATE.KL=((TARGET_STAFF-STAFF.K)/RECRUIT_DELAY)
/*   +A_LEAVING_RATE.K
D A_LEAVING_RATE=(PEOPLE/MONTH)    AVERAGE RATE AT WHICH
/*   PEOPLE LEAVE
D SMT=(MONTH) SMOOTHING TIME FOR AVERAGE
NOTE
```

```
NOTE  ****************************************************
RUN 4 SRL MODEL WITH PROPORTIONAL AND INERTIAL CONTROL
NOTE  ****************************************************
NOTE
R RECRUIT_RATE.KL=CLIP(ZERO,FIXED_RECRUITMENT_RATE,
/*   STAFF.K,TARGET_STAFF)
C FIXED_RECRUITMENT_RATE=5
D FIXED_RECRUITMENT_RATE=(PEOPLE/MONTH)    CONSTANT
/*   RECRUITMENT RATE
PRINT 3)RECRUIT_RATE
PRINT 4)LEAVING_RATE
C ZERO=0
D ZERO=(PEOPLE/MONTH)   ZERO
NOTE
NOTE   CLIP(A,B,C,D) FUNCTION IS EQUIVALENT TO THE
NOTE   IF...THEN STATEMENT IT SELECTS A IF C>=D AND
NOTE   B IF C<D (SEE APPENDIX 6)
NOTE
NOTE
NOTE  *************************************
RUN 5 SRL MODEL WITH FIXED RECRUITMENT RATE
NOTE  *************************************
```

Table 2 Model SRL2

```
NOTE   APPENDIX 1  TABLE 2     (MODEL SRL2)
NOTE
NOTE    DYSMAP2 EQUATIONS FOR RUNS 6 TO 11 OF THE
NOTE    STAFF RECRUITMENT AND LEAVING (SRL) MODEL
NOTE        OF CHAPTER 5
NOTE   (RESULTS IN FIGURES 5.7,5.9,5.11 AND 5.18 TO 5.23)
NOTE
NOTE   DIMENSIONLESS VARIABLES ARE GIVEN A DIMENSION OF (1)
NOTE
NOTE
L STAFF.K=STAFF.J+DT*(RECRUIT_RATE.JK-LEAVING_RATE.JK)
N STAFF=INITIAL_STAFF
R LEAVING_RATE.KL=STAFF.K*PROPORTION_LEAVING
C PROPORTION_LEAVING=0.05
R RECRUIT_RATE.KL=TABHL(POLTAB,TARGET_STAFF-STAFF.K,LO,
/*    HI,INC)
C LO=0
C HI=30
```

```
C INC=5
T POLTAB=0/2/3/3.5/4/4.5/5
C INITIAL_STAFF=0
C TARGET_STAFF=30
SPEC LENGTH=25/DT=0.1/PRTPER=1
PRINT 1)STAFF
PRINT 2)TARGET_STAFF
PRINT 3)RECRUIT_RATE
PRINT 4)LEAVING_RATE
N TIME=0
NOTE
NOTE    DOCUMENTATION OF VARIABLES
NOTE
D STAFF=(PEOPLE)     NUMBER OF CURRENT STAFF MEMBERS
D RECRUIT_RATE=(PEOPLE/MONTH)     RATE OF RECRUITMENT
/*     OF STAFF
D TARGET_STAFF=(PEOPLE)     TARGET NUMBER OF STAFF
D TIME=(MONTH)     SIMULATION TIME
D LENGTH=(MONTH)     SIMULATION LENGTH
D PRTPER=(MONTH)     SIMULATION PRINT PERIOD
D DT=(MONTH)     SIMULATION INTERVAL
D INITIAL_STAFF=(PEOPLE)     INITIAL NUMBER OF STAFF
D LEAVING_RATE=(PEOPLE/MONTH)     RATE OF LEAVING OF STAFF
D PROPORTION_LEAVING=(1/MONTH)     PROPORTION LEAVING
/*     PER MONTH
D POLTAB=(PEOPLE/MONTH)     POLICY TABLE FOR RECRUITMENT
/*     RATE
D LO=(PEOPLE)     LOWEST VALUE OF STAFF DESCREPANCY
/*     IN POLTAB
D HI=(PEOPLE)     HIGHEST VALUE OF STAFF DESCREPANCY
/*     IN POLTAB
D INC=(PEOPLE)     INCREMENT USED IN POLTAB
NOTE
NOTE ***********************************
RUN 6 SRL MODEL WITH NON-LINEAR CONTROL
NOTE ***********************************
NOTE
R RECRUIT_RATE.KL=NOR_RECRUIT_RATE*RECRUIT_MULTIPLIER
N NOR_RECRUIT_RATE=TARGET_STAFF*PROPORTION_LEAVING
A RECRUIT_MULTIPLIER.K=TABHL(POLTAB1,STAFF.K/
/*     TARGET_STAFF,LO,HI,INC)
C LO=0
C HI=1
```

```
C  INC=0.2
T  POLTAB1=5/4/3/2/1.5/1
D  RECRUIT_MULTIPLIER=(1)      MULTIPLIER FOR RECRUITMENT
/*    RATE
D  NOR_RECRUIT_RATE=(PEOPLE/MONTH)    NORMAL RECRUITMENT
/*    RATE
D  POLTAB1=(1)     POLICY TABLE FOR RECRUITMENT RATE
NOTE
NOTE *********************************
RUN 7 SRL MODEL WITH MULTIPLIER CONTROL
NOTE *********************************
NOTE
N  STAFF=TARGET_STAFF
A  NOR_RECRUIT_RATE.K=A_LEAVING_RATE.K
R  LEAVING_RATE.KL=PROPORTION_LEAVING*STAFF.K+STEP(HT,
/*    STIME)
C  STIME=2
C  HT=5
L  A_LEAVING_RATE.K=A_LEAVING_RATE.J+(DT/SMT)*(LEAVING
/*    _RATE.JK-A_LEAVING_RATE.J)
N  A_LEAVING_RATE=LEAVING_RATE
C  SMT=3
D  STIME=(MONTH)    TIME OF STEP INPUT TO LEAVING RATE
D  HT=(PEOPLE/MONTH)    HEIGHT OF STEP INPUT TO
/*    LEAVING RATE
D  A_LEAVING_RATE=(PEOPLE/MONTH)    AVERAGE RATE AT
/*    WHICH PEOPLE LEAVE
D  SMT=(MONTH) SMOOTHING TIME FOR AVERAGE
NOTE
NOTE *************************
RUN 8  FULLY BALANCED SRL MODEL
NOTE *************************
NOTE
L  STAFF.K=STAFF.J+DT*(TRAINING_END_RATE.JK-LEAVING_RATE.
/*    JK)
R  TRAINING_END_RATE.KL=DELAY3(RECRUIT_RATE.JK,TRAINING
/*    _DELAY)
C  TRAINING_DELAY=4
R  RECRUIT_RATE.KL=((TARGET_STAFF-STAFF.K)/RECRUIT_DELAY)
X  +A_LEAVING_RATE.K
C  RECRUIT_DELAY=5
L  STAFF_IN_TRAINING.K=STAFF_IN_TRAINING.J
X  +DT*(RECRUIT_RATE.JK-TRAINING_END_RATE.JK)
```

```
N STAFF_IN_TRAINING=A_LEAVING_RATE*TRAINING_DELAY
C LENGTH=40
PRINT 5)STAFF_IN_TRAINING
D TRAINING_END_RATE=(PEOPLE/MONTH) RATE AT WHICH
/*    TRAINING ENDS
D TRAINING_DELAY=(MONTH)    TRAINING DELAY
D RECRUIT_DELAY=(MONTH)     RECRUITMENT DELAY
D STAFF_IN_TRAINING=(PEOPLE)    STAFF IN TRAINING
NOTE
NOTE ***********************************************
RUN 9  SRL MODEL WITH DELAY AND PROPORTIONAL +INERTIAL
NOTE   CONTROL
NOTE ***********************************************
NOTE
R LEAVING_RATE.KL=BASE+STEP(HT,STIME)
N BASE=PROPORTION_LEAVING*STAFF
D BASE=(PEOPLE/MONTH)    BASE RATE FOR STAFF LEAVING
NOTE
NOTE ****************************************
RUN 10 SRL MODEL WITH EXOGENEOUS LEAVING RATE
NOTE ****************************************
NOTE
R RECRUIT_RATE.KL=((TARGET_STAFF-STAFF.K)/RECRUIT_DELAY)
X +A_LEAVING_RATE.K
X +((TARGET_STAFF_IN_TRAINING-STAFF_IN_TRAINING.K)/
/*    RECRUIT_DELAY)
A TARGET_STAFF_IN_TRAINING.K=A_LEAVING_RATE.K*TRAINING 1
/*    _DELAY
D TARGET_STAFF_IN_TRAINING=(PEOPLE)   TARGET STAFF IN
/*    TRAINING
NOTE
NOTE ***********************************************
RUN 11 SRL MODEL WITH DELAY AND PROPORTIONAL + INERTIAL
NOTE  + PIPELINE CONTROL
NOTE ***********************************************
```

Table 3 Model Delay B

```
NOTE
NOTE      APPENDIX 1    TABLE 3     (MODEL DELAYB)
NOTE
NOTE      DYSMAP2 EQUATIONS TO DEMONSTRATE THE
NOTE      RESPONSES OF FIRST, THIRD AND SIXTH ORDER
```

```
NOTE   DELAYS TO A PULSED INPUT (RESULTS IN FIGURE 5.16)
NOTE
R OUT.KL=DELAY1(IN.JK,DEL)*P1+DELAY3(IN.JK,DEL)*P2
X +DELAY3(IN1.JK,(DEL/2))*P3
R IN1.KL=DELAY3(IN.JK,(DEL/2))
R IN.KL=PULSE(5,5,40)
N IN=0
C DEL=10
C PRTPER=2
C LENGTH=30
C DT=.25
C P1=1
C P2=0
C P3=0
N TIME=0
RUN FIRST ORDER DELAY
C P2=1
C P1=0
C P3=0
RUN THIRD ORDER DELAY
C P3=1
C P1=0
C P2=0
RUN SIXTH ORDER DELAY
NOTE
NOTE   DOCUMENTATION
NOTE
D OUT=(UNITS/WEEK)   DELAY OUTFLOW RATE
D IN=(UNITS/WEEK)    DELAY INFLOW RATE
D IN1=(UNITS/WEEK)   SECOND DELAY INFLOW RATE
D DEL=(WEEK)      DELAY TIME
D P1=(1)      SWITCH FOR FIRST ORDER DELAY
D P2=(1)      SWITCH FOR THIRD ORDER DELAY
D P3=(1)      SWITCH FOR SIXTH ORDER DELAY
D LENGTH=(WEEK)      SIMULATION LENGTH
D TIME=(WEEK)       SIMULATION TIME
```

Appendix 2

DYSMAP2 Equations for the CIR Ltd Model of Chapter 6

(X or /* denotes a continuation line)

Table 1 Model CIRB

```
NOTE
NOTE        APPENDIX 2     MODEL(CIRB)
NOTE
NOTE       DYSMAP2 EQUATIONS FOR CIR LTD MODEL
NOTE    OF CHAPTER 6 (RESULTS IN FIGURES 6.6-6.10)
NOTE
*    CIR LIMITED
NOTE
NOTE ****************
NOTE PROFIT GENERATION
NOTE ****************
NOTE
A   AP.K=AR.K-AVC.K-AOH.K
A   POP.K=(AP.K/(AR.K+ZCR))*P
C   P=100
C   ZCR=.01
NOTE ZCR IS TO PREVENT DIVISION BY ZERO
NOTE
NOTE        ***********************
NOTE        GENERATION OF INVESTMENT
NOTE        ***********************
NOTE
R   IR.KL=((AP.K*PPE)/TTI)*ISW
```

```
C    PPE=0.5
C    ISW=0
C    TTI=12
NOTE
NOTE        **************************
NOTE        INVESTMENT IN SPACE SAVING
NOTE        **************************
NOTE
R    RISS.KL=DELAY3(IR.K,DEL)*RISSPPI.K
A    RISSPPI.K=TABHL(TCS,TIME.K,Z,ULT,IN)
C    ULT=120
C    IN=24
C    Z=0
T    TCS=.0005/.0005/.0005/.0005/.0005/.00016
L    SS.K=SS.J+DT*(RISS.JK-RRSS.JK)
N    SS=IPSS*ITS
C    IPSS=.45
C    ITS=166666.67
R    RRSS.KL=R.K*SSW
C    SSW=0
A    R.K=TABHL(TR,TIME.K,ZT,ULT,IN)
C    ZT=0
T    TR=700/700/500/500/300/300
L    SU.K=SU.J+DT*(-RISS.JK)
N    SU=(1-IPSS)*ITS
A    TSU.K=(SS.K+SU.K)
A    US.K=(SU.K/TSU.K)*100
R    CS.KL=TSU.K*UCS
C    UCS=1
NOTE
NOTE    ********************************
NOTE    INVESTMENT IN EMPLOYEE REDUCTION
NOTE    ********************************
NOTE
R    RODE.KL=(DELAY3(IR.K,DEL)*(CEPI.K/MI))
C    DEL=3
C    MI=1E6
A    CEPI.K=TABHL(TCEPI,TIME.K,ZT,ULT,IN)
T    TCEPI=30/25/20/20/10/10
L    E.K=E.J+DT*(-RODE.JK)
N    E=511
NOTE
NOTE            *********
```

```
NOTE            OVERHEADS
NOTE            * * * * * * * * *
NOTE
L    OH.K=OH.J+DT*(CS.JK+OOH.JK+OHIR.JK)
N    OH=IOH
C    IOH=8905000
R    OOH.KL=IOOH
C    IOOH=575334
R    OHIR.KL=INF*OH.K
C    INF=.0034
A    AOH.K=OOH1.K-OOH2.K
A    OHSF.K=OH.K-OOH1.K
L    OOH1.K=OOH1.J+DT*PULSE(OHSF.J/DT,IPT,PIN)
N    OOH1=OH
C    IPT=12
C    PIN=12
L    OOH2.K=OOH2.J+DT*PULSE((OOH1.J-OOH2.J)/DT,IPT,PIN)
N    OOH2=0
NOTE
NOTE        * * * * * * * * * * * * * *
NOTE            VARIABLE COSTS
NOTE        * * * * * * * * * * * * * *
NOTE
R    WC.KL=MWR.K*E.K
A    MWR.K=UWH*HPD*DPM
C    UWH=2
C    HPD=8
C    DPM=30
L    VC.K=VC.J+DT*(WC.JK+OVC.JK+VCIR.JK)
N    VC=4326000
R    VCIR.KL=INF*VC.K
R    OVC.KL=115000
A    AVC.K=OVC1.K-OVC2.K
A    VCSF.K=VC.K-OVC1.K
L    OVC1.K=OVC1.J+DT*PULSE(VCSF.J/DT,IPT,PIN)
N    OVC1=VC
L    OVC2.K=OVC2.J+DT*PULSE((OVC1.J-OVC2.J)/DT,IPT,PIN)
N    OVC2=0
NOTE
NOTE        * * * * * * * * * * * * * * * * *
NOTE            REVENUE GENERATION
NOTE        * * * * * * * * * * * * * * * * *
NOTE
```

```
L   REV.K=REV.J+DT*(REVIR.JK+RGR.JK)
N   REV=15763000
R   RGR.KL=1314000
A   RGT.K=TABHL(TVC,TIME.K,ZT,ULT,IN)
T   TVC=.0034/.0034/.0034/.0034/.0034/.0034
R   REVIR.KL=REV.K*RGT.K
A   AR.K=OREV1.K-OREV2.K
A   REVSF.K=REV.K-OREV1.K
L   OREV1.K=OREV1.J+DT*PULSE(REVSF.J/DT,IPT,
X   PIN)
N   OREV1=REV
L   OREV2.K=OREV2.J+DT*PULSE((OREV1.J-OREV2.J)/DT,IPT,PIN)
N   OREV2=0
NOTE
NOTE      *****************
NOTE      SIMULATION CONTROL
NOTE      *****************
NOTE
SPEC LENGTH=150/DT=.25/PRTPER=12
N   TIME=0
PRINT 1)POP,AP
PRINT 2)US,IR
PRINT 3)AVC,VC
PRINT 4)AOH,OH
PRINT 5)AR,REV,E
NOTE
NOTE ************
NOTE DOCUMENTATION
NOTE ************
NOTE
NOTE E=EMPLOYEES D=DAYS H=HOURS M=MONTH F=FEET P=POUND
NOTE
D   AR=(P) ANNUAL REVENUE
D   AP=(P) ANNUAL PROFIT
D   AVC=(P) ANNUAL VARIABLE COSTS
D   AOH=(P) ANNUAL OVERHEADS
D   CEPI=(E/P) CHANGE IN EMPLOYMENT PER MILLION
X   P INVESTED
D   CS=(P/M) COST OF SPACE PER MONTH
D   DEL=(M) INVESTMENT DELAY
D   DPM=(D/M) DAYS PER MONTH
D   DT=(M) SIMULATION INTERVAL
D   E=(E) EMPLOYEES
```

```
D   HPD=(H/D) HOURS PER DAY
D   IR=(P/M) INVESTMENT RATE
D   ISW=(1) INVESTMENT SWITCH
D   IN=(M) INTERVAL FOR TCEPI
D   IPT=(M) INITIAL PULSE TIME
D   PIN=(M) PULSE INTERVAL
D   INF=(1/M) INFLATION RATE
D   IOH=(P) INITIAL OVERHEADS
D   IPSS=(1) INITIAL % OF SPARE SPACE
D   IOOH=(P/M) INITIAL MONTHLY OTHER OVERHEAD COSTS
D   ITS=(F*F) INITIAL TOTAL SPACE
D   LENGTH=(M) SIMULATION LENGTH
D   MWR=(P/M/E) MONTHLY WAGE RATE
D   MI=(1) MILLION
D   OOH=(P/M) MONTHLY OTHER OVERHEAD COST RATE
D   OH=(P) OVERHEAD COSTS
D   OHIR=(P/M) OVERHEAD INCREASE RATE
D   OHSF=(P) OVERHEADS SO FAR IN YEAR
D   OVC=(P/M) OTHER VARIABLE COST
D   OVC1=(P) OLD VALUE OF VARIABLE COSTS 1
D   OVC2=(P) OLD VALUE OF VARIABLE COSTS 2
D   OOH1=(P) OLD VALUE OF OVERHEADS 1
D   OOH2=(P) OLD VALUE OF OVERHEADS 2
D   OREV1=(P) OLD VALUE OF REVENUE 1
D   OREV2=(P) OLD VALUE OF REVENUE 2
D   POP=(1) PERCENTAGE OPERATING PROFIT
D   P=(1) PERCENTAGE CONVERSION FACTOR
D   PPE=(1) PERCENTAGE OF PROFIT INVESTED
D   PRTPER=(M) SIMULATION PRINT PERIOD
D   REV=(P) REVENUE
D   REVIR=(P/M) REVENUE INCREASE RATE
D   REVSF=(P) REVENUE SO FAR IN YEAR
D   RGT=(1/M) RATE OF GROWTH IN TRADE
D   RGR=(P/M) REVENUE GENERATION RATE
D   RODE=(E/M) RATE OF DECREASE OF EMPLOYEES
D   RISS=((F*F)/M) RATE OF INCREASE IN SPARE SPACE
D   RRSS=((F*F)/M) RATE OF REDUCTION OF SPARE SPACE
D   RISSPPI=((F*F)/P) RATE OF INCREASE IN SPARE
X   SPACE PER POUND INVESTED
D   R=((F*F)/M) RATE OF REDUCTION OF SPARE
X   SPACE FOR TABLE
D   SSW=(1) SWITCH FOR SPACE REDUCTION STRATEGY
D   SS=(F*F) SPARE SPACE
```

```
D   SU=(F*F) SPACE USED
D   TCS=((F*F)/P) TABLE FOR RISSPPI
D   TTI=(M) TIME TAKEN FOR INVESTMENT
D   TR=((F*F)/M)  TABLE FOR ROS
D   TVC=(1/M) TABLE FOR APIV
D   TSU=(F*F) TOTAL COST OF SPACE
D   TCEPI=(E/P) TABLE FOR CEPI
D   TIME=(M) SIMULATION TIME
D   US=(1) % UTILISATION OF SPACE
D   UWH=(P/H/E) UNIT WAGE RATE
D   UCS=((P/M)/(F*F)) UNIT COST OF SPACE
D   ULT=(M) UPPER LIMIT OF TIME FOR TABLES
D   VC=(P) VARIABLE COST
D   VCSF=(P) VARIABLE COSTS SO FAR IN YEAR
D   VCIR=(P/M) VARIABLE COST INCREASE RATE
D   WC=(P/M) WAGES COST
D   ZCR=(P) RATIO CORRECTION FACTOR
D   Z=(P/M) ZERO RATE
D   ZT=(M) ZERO TIME
NOTE
NOTE  **************************
RUN   INFLATION=GROWTH IN REVENUE
NOTE  **************************
NOTE
T   TVC=.00234/.00216/.00216/.00208/.00200/.00196
NOTE
NOTE  ************************************
RUN   INFLATION EXCEEDS GROWTH IN REVENUE
NOTE  ************************************
NOTE
C   ISW=1
NOTE
NOTE  *********************
RUN   REDUCTION OF EMPLOYEES
NOTE  *********************
NOTE
C   SSW=1
NOTE
NOTE  **************************
RUN   ELIMINATION OF UNUSED SPACE
NOTE  **************************
NOTE
```

Appendix 3

STELLA Equations for the Coal Clearance Model of Chapter 7

Table 1 Equations for Policy I (Experiment 6) (Policy equations are in bold type)

bun1 = bun1 + dt * (bun_1_in_rate - bun_1_out_rate)
INIT(bun1) = 0

bun2 = bun2 + dt * (\bun_2_in_rate - bun_2_out_rate)
INIT(bun2) = 0

bun3 = bun3 + dt * (bun_3_in_rate - bun_3_out_rate)
INIT(bun3) = 0

cum_cf_rate = cum_cf_rate + dt * (tot_cf_rate)
INIT(cum_cf_rate) = 0

cum_coal_out_rates = cum_coal_out_rates + dt * (tot
    _coal_out_rate)
INIT(cum_coal_out_rates) = 0

bun_1_cap = 500

bun_1_in_rate = IF bun1 < 0.95*bun_1_cap THEN coalf1_out
    _rate ELSE 0

**bun_1_out_rate = IF (con_cap- cf_rate_bun2) >=
    max_out_rate_1 THEN   MIN(max_out_rate_1,
    MAX(0, bun1/DT)) ELSE 0**

bun_2_cap = 500

bun_2_in_rate = IF bun2 < .95* bun_2_cap THEN coalf2_out_
    rate ELSE 0

**bun_2_out_rate = IF (con_cap-cf_rate_bun3) >=
    max_out_rate_2 THEN MIN(max_out_rate_2,
    MAX(0 , bun2/DT)) ELSE 0**

bun_3_cap = 500

bun_3_in_rate = IF bun3 < 0.95*bun_3_cap THEN coalf3_
    out_rate ELSE 0

bun_3_out_rate = MIN(max_out_rate_3, MAX(0 ,bun3/DT))

cf_rate_bun1 = bun_1_out_rate+cf_rate_bun2

cf_rate_bun2 = bun_2_out_rate+cf_rate_bun3

cf_rate_bun3 = bun_3_out_rate

con_cap = 2500

efficiency = (cum_cf_rate/(cum_coal_out_rates +.1))*100

max_out_rate_1 = 1000

max_out_rate_2 = 1000

max_out_rate_3 = 1000

time_input = TIME

time_input_2 = TIME

time_input_3 = TIME

tot_cf_rate = cf_rate_bun1

tot_coal_out_rate = coalf1_out_rate+coalf2_out_rate+
    coalf3_out_rate

coalf1_out_rate = graph(time_input)
(0.0, 0.0),(0.251, 0.0),(0.503, 0.0),(0.754, 0.0),(1.01,

0.0),(1.26, 0.0),(1.51, 0.0),(1.76, 0.0),(2.01,
0.0),(2.26, 0.0),(2.51, 0.0),(2.76, 0.0),(3.02,
0.0),(3.27, 0.0),(3.52, 0.0),(3.77, 0.0),(4.02,
0.0),(4.27, 0.0),(4.52, 0.0),(4.77,0.0),(5.03,505.00),
(5.28,560.00),(5.53,600.00),(5.78,1357.50),(6.03,0.0),
(6.28,200.00),(6.53,1027.50),(6.79,1237.50),(7.04,
817.50),(7.29,99 0.00),(7.54,945.00),(7.79,1102.50),
(8.04,500.00),(8.29, 0.0),(8.54,997.50),(8.80,840.00),
(9.05,1275.00),(9.30,570.00),(9.55,1320.00),(9.80,
445.00),(10.05,1147.50),(10.30,450.00),(10.55,1395.00),
(10.81,750.00),(11.06,967.50),(11.31,595.00),(11.56,
0.0),(11.81,1267.50),(12.06, 0.0),(12.31, 0.0),(12.57,
0.0),(12.82, 0.0),(13.07, 0.0),(13.32,0.0),(13.57,
1342.50),(13.82,1177.50),(14.07,922.50),(14.32,
1320.00),(14.58,1027.50),(14.83,1155.00),(15.08,0.0),
(15.33,1230.00),(15.58,490.00),(15.83,1365.00),(16.08,
190.00),(16.34,995.00),(16.59,1282.50),(16.84,450.00),
(17.09,0.0),(17.34,1185.00),(17.59,795.00),(17.84,
1207.50),(18.09,877.50),(18.35,997.50),(18.60,1102.50),
(18.85,952.50),(19.10, 0.0),(19.35,1125.00),(19.60,
1012.50),(19.85, 0.0),(20.10, 0.0),(20.36, 0.0),
(20.61, 0.0),(20.86, 0.0),(21.11, 0.0),(21.36, 0.0),
(21.61, 0.0),(21.86, 0.0),(22.12, 0.0),(22.37, 0.0),
(22.62, 0.0),(22.87, 0.0),(23.12, 0.0),(23.37, 0.0),
(23.62, 0.0),(23.87, 0.0),(24.13, 0.0),(24.38, 0.0),
(24.63, 0.0),(24.88, 0.0),(25.13, 0.0)...

coalf2_out_rate = graph(time_input_2)
(0.0, 0.0),(0.251, 0.0),(0.503, 0.0),(0.754, 0.0),
(1.01, 0.0),(1.26, 0.0),(1.51, 0.0),(1.76, 0.0),(2.01,
0.0),(2.26, 0.0),(2.51, 0.0),(2.76, 0.0),(3.02, 0.0),
(3.27, 0.0),(3.52, 0.0),(3.77, 0.0),(4.02, 0.0),(4.27,
0.0),(4.52, 0.0),(4.77,0.0),(5.03,1350.00),(5.28,
1102.50),(5.53,982.50),(5.78,1267.50), (6.03,0.0),
(6.28,1380.00),(6.53,915.00),(6.79,1185.00),(7.04,
1050.00),(7.29,1192.50),(7.54,1222.50),(7.79,1117.50),
(8.04,1260.00),(8.29, 0.0),(8.54,1327.50),(8.80,952.50),
(9.05,1162.50),(9.30,1005.00),(9.55,112 5.00),(9.80,
870.00),(10.05,1132.50),(10.30,997.50),(10.55,735.00),
(10.81, 1237.50),(11.06,600.00),(11.31,1260.00),(11.56,
0.0),(11.81,1440.00),(12.06, 0.0),(12.31, 0.0),(12.57,
0.0),(12.82, 0.0),(13.07, 0.0),(13.32, 0.0),(13.57,
1410.00),(13.82,200.00),(14.07,1380.00),(14.32,

1297.50),(14.58,1222.50),(14.83,1192.50),(15.08,
1185.00),(15.33,1252.50),(15.58,1035.00),(15.83,
1087.50),(16.08,1117.50),(16.34,995.00),(16.59,
1065.00),(16.84,1057.50),(17.09,0.0),(17.34,1147.50),
(17.59,982.50),(17.84,1117.50),(18.09,1230.00),(18.35,
1267.50),(18.60,1222.50),(18.85,1260.00),(19.10,
0.0),(19.35,450.00),(19.60,1177.50),(19.85, 0.0),(20.10,
0.0),(20.36, 0.0),(20.61, 0.0),(20.86, 0.0),(21.11,
0.0),(21.36, 0.0),(21.61, 0.0),(21.86, 0.0),(22.12,
0.0),(22.37, 0.0),(22.62, 0.0),(22.87, 0.0),(23.12,
0.0),(23.37, 0.0),(23.62, 0.0),(23.87, 0.0),(24.13,
0.0),(24.38, 0.0),(24.63, 0.0),(24.88, 0.0),(25.13,
0.0)...

coalf3_out_rate = graph(time_input_3)
(0.0, 0.0),(0.251, 0.0),(0.503, 0.0),(0.754, 0.0),(1.01,
0.0),(1.26, 0.0),(1.51, 0.0),(1.76, 0.0),(2.01, 0.0),
(2.26, 0.0),(2.51, 0.0),(2.76, 0.0),(3.02, 0.0),(3.27,
0.0),(3.52, 0.0),(3.77, 0.0),(4.02, 0.0),(4.27, 0.0),(4.52,
0.0),(4.77,0.0),(5.03,1200.00),(5.28,1275.00),(5.53,
1192.50),(5.78,1252.50),(6.03,1215.00),(6.28,1260.00),
(6.53,1147.50),(6.79,1102.50),(7.04,885.00),(7.29,
300.00),(7.54,1365.00),(7.79,1380.00),(8.04,1200.00),
(8.29, 0.0),(8.54,1320.00),(8.80,1222.50),(9.05,
1305.00),(9.30,450.00),(9.55,115 5.00),(9.80,1102.50),
(10.05,1185.00),(10.30,1282.50),(10.55,1282.50),(10.
81,1192.50),(11.06,1185.00),(11.31,595.00),(11.56,0.0),
(11.81,1327.50),(12.06, 0.0),(12.31, 0.0),(12.57, 0.0),
(12.82, 0.0),(13.07, 0.0),(13.32, 0.0),(13.57,1207.50),
(13.82,1245.00),(14.07,1162.50),(14.32,1327.50),(14.58,
1260.00),(14.83,1335.00),(15.08,1297.50),(15.33,
1320.00),(15.58,490.00),(15.83,1267.50),(16.08,1140.00),
(16.34,995.00),(16.59,1050.00),(16.84,450.00),(17.09,
967.50),(17.34,450.00),(17.59,1230.00),(17.84,
1350.00),(18.09,1222.50),(18.35,1327.50),(18.60,370.00),
(18.85,1132.50),(19.10, 0.0),(19.35,1125.00),(19.60,
1230.00),(19.85, 0.0),(20.10, 0.0),(20.36, 0.0),(20.61,
0.0),(20.86, 0.0),(21.11, 0.0),(21.36, 0.0),(21.61,
0.0),(21.86, 0.0),(22.12, 0.0),(22.37, 0.0),(22.62,
0.0),(22.87, 0.0),(23.12, 0.0),(23.37, 0.0),(23.62,
0.0),(23.87, 0.0),(24.13, 0.0),(24.38, 0.0),(24.63,
0.0),(24.88, 0.0),(25.13, 0.0)...

Table 2 Equations for Policy II (Experiment 6) (Policy equations are in bold type)

```
bun1 = bun1 + dt * ( bun_1_in_rate - bun_1_out_rate )
INIT(bun1) = 0


bun2 = bun2 + dt * ( bun_2_in_rate - bun_2_out_rate )
INIT(bun2) = 0


bun3 = bun3 + dt * ( bun_3_in_rate - bun_3_out_rate )
INIT(bun3) = 0


cum_cf_rate = cum_cf_rate + dt * ( tot_cf_rate )
INIT(cum_cf_rate) = 0


cum_coal_out_rates = cum_coal_out_rates + dt * ( tot_
    coal_out_rate )
INIT(cum_coal_out_rates) = 0


bun_1_cap = 500


bun_1_in_rate = IF bun1 < 0.95*bun_1_cap THEN coalf1_
    out_rate ELSE 0
```

bun_1_out_rate = IF (con_cap- cf_rate_bun2) >=
des_out_rate_1 THEN   MIN(des_out_rate_1,
MAX(0, bun1/DT)) ELSE
MIN(con_cap-cf_rate_bun2,MAX(0,bun1/DT))

```
bun_2_cap = 500


bun_2_in_rate = IF bun2 < .95* bun_2_cap THEN coalf2_
out_rate ELSE 0
```

bun_2_out_rate =   IF (con_cap-cf_rate_bun3) >=
des_out_rate_2 THEN MIN(des_out_rate_2,
MAX(0 , bun2/DT)) ELSE
MIN(con_cap-cf_rate_bun3,MAX(0,bun2/DT))

```
bun_3_cap = 500


bun_3_in_rate = IF bun3 < 0.95*bun_3_cap THEN coalf3_out_
    rate ELSE 0
```

```
bun_3_out_rate = MIN(des_out_rate_3, MAX(0 ,bun3/DT))

cf_rate_bun1 = bun_1_out_rate+cf_rate_bun2

cf_rate_bun2 = bun_2_out_rate+cf_rate_bun3

cf_rate_bun3 = bun_3_out_rate

con_cap = 2500

des_out_rate_1 = (bun1*max_out_rate_1)/bun_1_cap

des_out_rate_2 = (bun2*max_out_rate_2)/bun_2_cap

des_out_rate_3 = (bun3*max_out_rate_3)/bun_3_cap

efficiency = (cum_cf_rate/(cum_coal_out_rates +.1))*100

max_out_rate_1 = 1000

max_out_rate_2 = 1000

max_out_rate_3 = 1000

time_input = TIME

time_input_2 = TIME

time_input_3 = TIME

tot_cf_rate = cf_rate_bun1

tot_coal_out_rate = coalf1_out_rate+coalf2_out_rate+
    coalf3_out_rate

coalf1_out_rate = graph(time_input)  _ as for policy1

coalf2_out_rate = graph(time_input_2) _ as for policy1

coalf3_out_rate = graph(time_input_3) _ as for policy1
```

Table 3 Equations for Policy III (Experiment 6) (Policy equations are in bold type)

```
bun1 = bun1 + dt * ( bun_1_in_rate - bun_1_out_rate )
INIT(bun1) = 0

bun2 = bun2 + dt * ( bun_2_in_rate - bun_2_out_rate )
INIT(bun2) = 0

bun3 = bun3 + dt * ( bun_3_in_rate - bun_3_out_rate )
INIT(bun3) = 0

cum_cf_rate = cum_cf_rate + dt * ( tot_cf_rate )
INIT(cum_cf_rate) = 0

cum_coal_out_rates = cum_coal_out_rates + dt * ( tot_
coal_out_rate )
INIT(cum_coal_out_rates) = 0
```

bun1_sat_sw = IF bun1 > saturation_level THEN 1 ELSE 0

bun2_sat_sw = IF bun2 > saturation_level THEN 1 ELSE 0

bun3_sat_sw = IF bun3 > saturation_level THEN 1 ELSE 0

```
bun_1_in_rate =IF bun1< 0.95*bun_cap THEN coalf1_
out_rate ELSE 0
```

bun_1_out_rate =IF no_of_buns_saturate >=2 THEN
MIN(MIN(con_cap/3 , max_out_rate_1),MAX(0,bun1/DT))
ELSE IF bun1 >= saturation_level THEN MIN(max_out_rate_1,
MAX(0,bun1/DT)) ELSE MIN ((bun1/(sum_buns_non_sat+.1))
**   *new_con_cap_left ,**
MAX(0,bun1/DT))

```
bun_2_in_rate = IF bun2 < .95* bun_cap THEN coalf2_out_
rate ELSE 0
```

bun_2_out_rate = IF no_of_buns_saturate >= 2 THEN
MIN(MIN(con_cap/3 ,max_out_rate_2),MAX(0,bun2/DT))
ELSE IF bun2 >=saturation_level THEN MIN(max_out_rate_2,
MAX(0,bun2/DT))
ELSE MIN((bun2/(sum_buns_non_sat+.1))*new_con_cap_left ,

```
MAX(0,bun2/DT))
```

bun_3_in_rate = IF bun3 < 0.95*bun_cap THEN coalf3_out_
rate ELSE 0

bun_3_out_rate = IF no_of_buns_saturate >= 2 THEN
MIN(MIN(con_cap/3 ,max_out_rate_3),MAX(0,bun3/DT))
ELSE IF bun3 >=saturation_level THEN MIN(max_out_rate_3 ,
MAX(0,bun3/DT))
ELSE MIN((bun3/(sum_buns_non_sat+.1))*new_con_cap_left ,
MAX(0,bun3/DT))

bun_cap = 500

cf_rate_bun1 = bun_1_out_rate+cf_rate_bun2

cf_rate_bun2 = bun_2_out_rate+cf_rate_bun3

cf_rate_bun3 = bun_3_out_rate

con_cap = 2500

con_cap_left =MAX(0,
con_cap-(no_of_full_buns*max_out_rate_1))

efficiency = (cum_cf_rate/(cum_coal_out_rates +.1))*100

max_out_rate_1 = 1000

max_out_rate_2 = 1000

max_out_rate_3 = 1000

new_con_cap_left =
con_cap-(no_of_buns_saturate*max_out_rate_1)

no_of_buns_saturate =
bun3_sat_sw+bun2_sat_sw+bun1_sat_sw

no_of_full_buns =
INT((bun1+.1)/bun_cap)+INT((bun2+.1)/bun_cap)+INT
((bun3+.1)/bun_cap)

```
saturation_level =
(max_out_rate_1/(con_cap_left+.1))*sum_levs_unful_buns

sum_buns = bun1+bun2+bun3

sum_buns_non_sat =
MIN(bun1,saturation_level)+MIN(bun2,saturation_level)+
  MIN(bun3,saturation_level)-(no_of_buns_saturate
  *saturation_level)

sum_levs_unful_buns = sum_buns-(no_of_full_buns*bun_cap)
```

time_input = TIME

time_input_2 = TIME

time_input_3 = TIME

tot_cf_rate = cf_rate_bun1

tot_coal_out_rate = coalf1_out_rate+coalf2_out_rate+
    coalf3_out_rate

coalf1_out_rate = graph(time_input)   _as for policy I

coalf2_out_rate = graph(time_input_2) as for policy I

coalf3_out_rate = graph(time_input_3) as for policy I

Appendix 4

(X or /* denotes a continuation line)

Table 1 DYSMAP2 Equations for the Armoured Advance Model of Chapter 8

```
NOTE         APPENDIX 4      TABLE 1
NOTE
NOTE      DYSMAP2 EQUATIONS FOR THE
NOTE   ARMOURED ADVANCE MODEL OF CHAPTER 8
NOTE ********************************************
NOTE * *
*     *        ARMOURED ADVANCE   MODEL *
NOTE *     (FILE FAAMB)    *
NOTE * *
NOTE ********************************************
NOTE
NOTE VARIABLES TO BE CHANGED FOR EXPERIMENTS
NOTE          IN THIS BOOK
NOTE
C  P1=0
C  DS=0
C  SS=0
C  MS=1
C  PWDRA=1500
C  PWDRG=1500
NOTE
NOTE ********************************************
NOTE *
NOTE *    ADVANCE IN BATTALION FORMATION *
NOTE *
NOTE ********************************************
NOTE
NOTE          ** RED  MOVEMENTS **
NOTE
R  ASR.KL=MAX(ZAODR,UIAA.K/DT)
C  ZAODR=0
```

```
A   WDAS.K=CLIP(0,1,ZFR,WDRA.K)
C   ZFR=0
A   BFS.K=CLIP(1,0,NABF.K,MNA)*(1-OARS.K)
C   MNA=1
L   UIAA.K=UIAA.J+DT*(-ASR.JK)
N   UIAA=UIAAX
C   UIAAX=1800
R   CCDR.KL=MAX(ZAODR,NABF.K/DT)*
X   CLIP(1,0,ADBF.K,PDCCD.K)
L   NABF.K=NABF.J+DT*(ASR.JK-CCDR.JK-ARBF.JK)
N   NABF=0
NOTE
NOTE ** DISTANCE AND SPEED CALCULATIONS   **
NOTE
A   PDCCD.K=P1*FDCCD+(1-P1)*VDCCD.K
C   FDCCD=160
A   VDCCD.K=(MIN(FDCCD*BFSM.K,TD-1)*SZS
X   +MIN(FDCCD*BFZM.K,TD-1)*(1-SZS))
X   *(1-ARS.K)
C   SZS=0.5
A   BFZM.K=TABHL(BFZMT,BFZR.K,0,1.0,.20)
T   BFZMT=0/.55/.75/.90/.97/1.0
A   BFZR.K=(NABF.K/UIAAX)*BFS.K
A   BFSM.K=TABHL(BFSMT,BFSR.K,0,1.0,.25)
T   BFSMT=1.5/1.28/1.18/1.09/1.0
A   BFSR.K=(ASBF.K/PSBF)*BFS.K
A   AMBF.K=ASBF.K*NABF.K
C   PSBF=40
L   ADBF.K=ADBF.J+DT*(RCDBF.JK)
N   ADBF=0
NOTE
NOTE **  BLUE  WEAPON DELIVERY CALCULATIONS   **
NOTE
A   WDRAD.K=PWDRA*BFS.K
X   *(CLIP(1,0,ADBF.K,ZD)
X   -CLIP(1,0,ADBF.K,(FDCCD*PD1))
X   +CLIP(1,0,ADBF.K,(FDCCD*PD2))
X   -CLIP(1,0,ADBF.K,(FDCCD*PD22)))
C   ZD=0
C   PD1=.2
C   PD22=0.9
A   WDRAS.K=PWDRA*BFS.K*FLSB.K
L   OSBF.K=OSBF.J+DT*(ASBF.J-OSBF.J)/DT
```

```
N    OSBF=0
A    FLSB.K=CLIP(1,0,ASBF.K,BFSLL)
X    *CLIP(1,CLIP(0,1,ASBF.K,OSBF.K),ASBF.K,BFSUL)
A    WDRAM.K=PWDRA*BFS.K*FLMB.K
L    OMBF.K=OMBF.J+DT*(AMBF.J-OMBF.J)/DT
N    OMBF=0
A    FLMB.K=CLIP(1,0,AMBF.K,BFMLL)
X    *CLIP(1,CLIP(0,1,AMBF.K,OMBF.K),AMBF.K,BFMUL)
C    BFSLL=18
C    BFSUL=30
C    BFMLL=25000
C    BFMUL=30000
A    DWDRA.K=WDRAD.K*DS+WDRAS.K*SS
X    +WDRAM.K*MS
A    WDRA.K=MIN(DWDRA.K,MAX(ZFR,(ASA-CAA.K)/DT))
C    ASA=20000
C    PD2=.8
NOTE
NOTE    ** EFFECTS OF BLUE FIRE   **
NOTE
R    RDSBF.KL=MIN((WDRA.K*PPS.K),MAX(ZR,ASBF.K/DT))
X    *BFS.K*WDAS.K
C    ZR=0
R    RISBF.KL=((PSBF-ASBF.K)/TCS.K)*(1-WDAS.K)
A    TCS.K=TABHL(ST,PSBF-ASBF.K,0,40,10)
T    ST=.15/.15/.25/.5/1
L    NSBF.K=NSBF.J+DT*(-RDSBF.JK+RISBF.JK)
N    NSBF=PSBF
A    ASBF.K=MAX(NSBF.K,BS)*(1-OARS.K)
R    RCDBF.KL=(MIN(ASBF.K,MAX(ZS,(PDCCD.K-ADBF.K)/DT)))
X    *BFS.K*P1
X    +ASBF.K*(1-P1)*BFS.K*CLIP(0,1,ZD,DTE.K)
C    ZS=0
A    CCDR1.K=CCDR.KL
R    ARBF.KL=MIN((WDRA.K*PPSA.K),MAX(ZAODR,
X    ((NABF.K/DT)-CCDR1.K)))*BFS.K
X    *WDAS.K
NOTE
NOTE    ****************************************************
NOTE    * *
NOTE    *    ADVANCE IN  COMPANY COLUMNS *
NOTE    * *
NOTE    ****************************************************
```

```
NOTE
NOTE   **     RED    MOVEMENTS    **
NOTE
A   CCS.K=CLIP(1,0,NACC.K,MNA)*(1-OARS.K)
A   WDGS.K=CLIP(0,1,ZFR,WDRG.K)
L   NACC.K=NACC.J+DT*(CCDR.JK-PCDR.JK-ARCC.JK)
N   NACC=0
R   PCDR.KL=MAX(ZAODR,NACC.K/DT)*CLIP(1,0,ADCC.K,PDPCD.K)
NOTE
NOTE   ** DISTANCE AND SPEED CALCULATIONS **
NOTE
A   PDPCD.K=P1*FDPCD+(1-P1)*MDPCD.K
C   FDPCD=80
A   MDPCD.K=TD-ADBF.K
NOTE
NOTE   ** BLUE WEAPON DELIVERY CALCULATIONS **
NOTE
A   WDRGD.K=PWDRG*CCS.K
X   *(CLIP(1,0,ADCC.K,ZD)
X   -CLIP(1,0,ADCC.K,PDPCD.K*PD3)
X   +CLIP(1,0,ADCC.K,PDPCD.K*PD4)
X   -CLIP(1,0,ADCC.K,PDPCD.K*PD44))
C   PD3=.3
C   PD44=.9
A   WDRGS.K=PWDRG*CCS.K*FLSC.K
L   OSCC.K=OSCC.J+DT*(ASCC.J-OSCC.J)/DT
N   OSCC=0
A   WDRGM.K=PWDRG*CCS.K*FLMC.K
L   OMCC.K=OMCC.J+DT*(AMCC.J-OMCC.J)/DT
N   OMCC=0
A   FLSC.K=CLIP(1,0,ASCC.K,CCSLL)
X   *CLIP(1,CLIP(0,1,ASCC.K,OSCC.K),ASCC.K,CCSUL)
A   FLMC.K=CLIP(1,0,AMCC.K,CCMLL)
X   *CLIP(1,CLIP(0,1,AMCC.K,OMCC.K),AMCC.K,CCMUL)
C   CCSLL=8
C   CCSUL=12
C   CCMLL=6000
C   CCMUL=12000
A   AMCC.K=ASCC.K*NACC.K
A   DWDRG.K=WDRGD.K*DS+WDRGS.K*SS
X   +WDRGM.K*MS
A   WDRG.K=MIN(DWDRG.K,MAX(ZFR,(ASG-CAG.K)/DT))
C   ASG=100000
```

```
C   PD4=.7
NOTE
NOTE   ** EFFECTS OF BLUE FIRE   **
NOTE
R   RDSCC.KL=MIN((WDRG.K*PPS.K),MAX(ASCC.K/DT,ZR))
X   *CCS.K*WDGS.K
L   NSCC.K=NSCC.J+DT*(-RDSCC.JK+RISCC.JK)
N   NSCC=PSCC
A   ASCC.K=MAX(NSCC.K,BS)*(1-OARS.K)
C   BS=5
R   RISCC.KL=((PSCC-ASCC.K)/TCSC)*(1-WDGS.K)*CCS.K
C   TCSC=0.25
C   PSCC=15
R   RCDCC.KL=((MIN(ASCC.K,MAX(ZS,(PDPCD.K-ADCC.K)/DT))))
X   *CCS.K*P1+ASCC.K*(1-P1)*CCS.K*CLIP(0,1,D,DTE.K))
C   D=0
L   ADCC.K=ADCC.J+DT*(RCDCC.JK)
N   ADCC=0
A   PCDR1.K=PCDR.KL
R   ARCC.KL=MIN((WDRG.K*PPSA.K),MAX(ZAODR,
X   ((NACC.K/DT)-PCDR1.K)))*CCS.K*WDGS.K
NOTE
NOTE   **********************************************
NOTE   *   *
NOTE   *          GENERAL EQUATIONS   *
NOTE   *   *
NOTE   **********************************************
NOTE
NOTE   **      CALCULATION OF DISTANCES   **
NOTE
A   CD.K=INT(ADBF.K+ADCC.K)
A   DTE.K=INT(TD-CD.K)
C   TD=240
L   DDTE.K=DDTE.J+(DT/PD)*(DTE.J-DDTE.J)
N   DDTE=TD
C   PD=0.25
A   PDTE.K=DDTE.K*P
C   P=0.9
NOTE
NOTE   **  CALCULATION OF FIRE PRODUCTIVITIES   **
NOTE
A   PPSA.K=NARPS*WAM.K*TDF.K
C   NARPS=0.015
```

```
A   PPS.K=NSRPS*WAM.K*TDF.K
C   NSRPS=0.0015
A   WAM.K=TABHL(WAT,PDTE.K,0,240,40)*(1-ARS.K)
T   WAT=2/1.5/.5/.5/.5/.4/.2
A   TDF.K=CLIP(1,0,CCS.K,1)+CLIP(15,0,BFS.K,1)
NOTE
NOTE **  CALCULATION OF AMMUNITION USEAGE   **
NOTE
L   CAA.K=CAA.J+DT*(WDRA.J)
N   CAA=0
L   CAG.K=CAG.J+DT*(WDRG.J)
N   CAG=0
A   CA.K=CAA.K+CAG.K
NOTE
NOTE  ** CALCULATION OF ARRIVAL TIME   **
NOTE
A   ARS.K=CLIP(1,0,CD.K,TD)
X   +CLIP(1,CLIP(1,0,TIME.K,LENGTH),CAR.K,UIAAX-1)
X   *CLIP(0,1,CD.K,TD)
L   OARS.K=OARS.J+DT*((ARS.J-OARS.J)/DT)
N   OARS=0
L   ART.K=ART.J+DT*ARS.J*(1-OARS.J)*(TIME.J/DT)
N   ART=0
NOTE
NOTE  ** CALCULATION OF ARRIVAL SIZE   **
NOTE
A   NA.K=NABF.K*BFS.K+NACC.K*CCS.K
L   ARSZ.K=ARSZ.J+DT*ARS.J*(1-OARS.J)*(NA.J/DT)
N   ARSZ=0
NOTE
NOTE ** CALCULATION OF TIME TO COMPANY COLUMN   **
NOTE     DEPLOYMENT
NOTE
L   OCCS.K=OCCS.J+DT*((CCS.J-OCCS.J)/DT)
N   OCCS=0
L   TTCCD.K=TTCCD.J+DT*CCS.J*(1-OCCS.J)*(TIME.J/DT)
N   TTCCD=0
NOTE
NOTE  **  CALCULATION OF ARRIVAL MOMENTUM   **
NOTE
A   MOM.K=(ARSZ.K/(ART.K+CLIP(H,STT.K-ART.K,
X   ART.K,STT.K)))*ARS.K
C   H=0
```

```
L   CCDS.K=CCDS.J+DT*CCS.J*(1-OCCS.J)*(NA.J/DT)
N   CCDS=0
NOTE
NOTE **  CALCULATE AVERAGE REDUCTION IN MOMENTUM   **
NOTE PER 1000 SHELLS FIRED
NOTE
A   STT.K=(FDCCD/PSBF)+(FDPCD/PSCC)
A   ARMPKS.K=(((UIAAX/(STT.K+SWITCH(MT,ZT,STT.K)))-MOM
X   .K)/(CA.K+SWITCH(MA,ZA,CA.K)))*ARS.K*TH
C   MT=1
C   MA=1
C   ZA=0
C   ZT=0
C   TH=1000
NOTE
NOTE   **    CALCULATION OF CUMMULATIVE ATTRITION   **
NOTE
L   CAR.K=CAR.J+DT*(ARBF.JK*BFS.J+ARCC.JK*CCS.J)
N   CAR=0
A   NAP.K=NABF.K*BFS.K+NACC.K*CCS.K
A   SPP.K=ASBF.K*BFS.K+ASCC.K*CCS.K
NOTE
NOTE ******************************************
NOTE * CONTROL STATEMENTS        *
NOTE ******************************************
NOTE
C   DT=0.0625
C   PRTPER=20
C   LENGTH=20
N   TIME=0
PRINT 1)TTCCD,ARS,NAP,SPP
PRINT 2)ART,CCS,ADCC
PRINT 3)ARSZ,WDRAD,WDRAS
PRINT 4)MOM,NACC,WDRAM
PRINT 5)CAR,CD,NA,CCDS
PRINT 6)ARMPKS,BFS,NABF
PRINT 7)CA,CAA,CAG,SZS
PRINT 8)WDRGD,WDRGS,WDRGM
NOTE ******************************************
NOTE *    *
NOTE *     DOCUMENTATION  *
NOTE *    *
NOTE ******************************************
```

```
NOTE
NOTE      ABBREVIATIONS
NOTE    K=KILOMETERS H=HOURS V=VEHICLES S=SHELLS
NOTE
D   ADCC=(K)      ACTUAL DISTANCE IN COMPANY COLUMNS
D   ADBF=(K)      ACTUAL DISTANCE IN BATTALION FORMATION
D   AMBF=(V*(K/H))   ACTUAL MOMENTUM IN BATTALION
/* FORMATION
D   AMCC=(V*(K/H))   ACTUAL MOMENTUM IN COMPANY COLUMN
D   ARBF=(V/H)     ATTRITION RATE IN BATTALION FORMATION
D   ARCC=(V/H)     ATTRITION RATE IN COMPANY COLUMN
D   ARMPKS=((V/H)/S) AVERAGE REDUCTION IN MOMENTUM PER
/* 1000 SHELLS
D   ARS=(1)      SWITCH TO MARK ARRIVAL OF RED AT BLUE'S
/* POSITION
D   ARSZ=(V)     SIZE OF RED ARRIVAL FORCE AT BLUE
/* POSITION
D   ART=(H)      TIME OF RED ARRIVAL AT BLUE POSITION
D   ASBF=(K/H)     ACTUAL SPEED IN BATTALION FORMATION
D   ASG=(S)     AVAILABLE SHELLS ARTILLERY
D   ASA=(S)     AVAILABLE SHELLS AIRCRAFT
D   ASCC=(K/H)     ACTUAL SPEED IN COMPANY COLUMNS
D   ASR=(V/H)     ADVANCE START RATE
D   BFMLL=(V*(K/H))  BATTALION FORMATION MOMENTUM LOWER
/*  LIMIT
D   BFMUL=(V*(K/H))  BATTALION FORMATION MOMENTUM UPPER
/*  LIMIT
D   BFSLL=(K/H)     BATTALION FORMATION SPEED LOWER LIMIT
D   BFSUL=(K/H)     BATTALION FORMATION SPEED UPPER LIMIT
D   BFS=(1)     BATTALION FORMATION SWITCH
D   BFSM=(1)     BATTALION FORMATION SPEED MULTIPLIER
D   BFSMT=(1)     BATTALION FORMATION SPEED TABLE
D   BFSR=(1)     BATTALION FORMATION SPEED RATIO
D   BFZM=(1)     BATTALION FORMATION SIZE MULTIPLIER
D   BFZMT=(1)     BATTALION FORMATION SIZE TABLE
D   BFZR=(1)     BATTALION FORMATION SIZE RATIO
D   BS=(K/H)     BASE SPEED
D   CAA=(S)     CUMULATIVE AMMUNITION USED BY AIRCRAFT
D   CA=(S)     CUMULATIVE AMMUNITION
D   CAR=(V)     CUMULATIVE ATTRITION RATE
D   CCMLL=(V*(K/H))  COMPANY COLUMN MOMENTUM LOWER LIMIT
D   CCMUL=(V*(K/H))  COMPANY COLUMN MOMENTUM UPPER LIMIT
D   CCSLL=(K/H)     COMPANY COLUMN SPEED LOWER LIMIT
```

```
D   CCSUL=(K/H)      COMPANY COLUMN SPEED UPERR LIMIT
D   CCDS=(V)      SIZE OF FORCE AT COMPANY COLUMN DEPLOYMENT
D   CAG=(S)      CUMULATIVE AMMUNITION USED BY ARTILLERY
D   CCS=(1)      COMPANY COLUMN SWITCH
D   CCDR=(V/H)      COMPANY COLUMN DEPLOYMENT RATE
D   CD=(K)      CUMULATIVE DISTANCE
D   CCDR1=(V/H)      COMPANY COLUMN DEPLOYMENT RATE 1
D   D=(K)      ZERO DISTANCE
D   DDTE=(K)      DELAYED DISTANCE TO ENEMY
D   DS=(1)      DISTANCE CRITERION SWITCH
D   DT=(H)      SIMULATION INTERVAL
D   DTE=(K)      DISTANCE TO ENEMY
D   DWDRA=(S/H)      DESIRED WEAPON DELIVERY RATE
/*   AIRCRAFT
D   DWDRG=(S/H)      DESIRED WEAPON DELIVERY RATE
/*   ARTILLERY
D   FDPCD=(K)      FIXED DISTANCE TO PLATOON COLUMN
/*   DEPLOYMENT
D   FDCCD=(K)      FIXED DISTANCE TO COMPANY COLUMN
/*   DEPLOYMENT
D   FLMB=(1)      FLIP FUNCTION FOR FIRE DELIVERY ON
/*   BATTALION FORMATION ON MOMENTUM CRITERION
D   FLMC=(1)      FLIP FUNCTION FOR FIRE DELIVERY ON
/*   COMPANY COLUMN FORMATION ON MOMENTUM CRITERION
D   FLSB=(1)      FLIP FUNCTION FOR FIRE DELIVERY ON
/*   BATTALION FORMATION ON SPEED CRITERION
D   FLSC=(1)      FLIP FUNTION FOR FIRE DELIVERY ON
/*   COMPANY COLUMN FORMATION ON SPEED CRITERION
D   H=(H)      ZERO TIME
D   MOM=(V/H)      RED MOMENTUM ON ARRIVAL
D   LENGTH=(H)      SIMULATION LENGTH
D   MNA=(V)      MINIMUM NUMBER ADVANCING
D   MDPCD=(K)      MODIFIED DISTANCE TO PLATOON COLUMN
/*   DEPLOYMENT
D   MT=(H)      MINIMUM TIME
D   MA=(S)      MINIMUM AMMUNITION
D   MS=(1)      MOMENTUM CRITERION SWITCH
D   NABF=(V)      NUMBER ADVANCING IN BATTALION FORMATION
D   NACC=(V)      NUMBER ADVANCING IN COMPANY COLUMNS
D   NSCC=(K/H)      NORMAL SPEED IN COMPANY COLUMN
D   NSBF=(K/H)      NORMAL SPEED IN BATTALION FORMATION
D   NSRPS=((K/H)/S)   NORMAL SPEED REDUCTION PER SHELL
D   NA=(V)      NUMBER ADVANCING (FOR PLOTTING)
```

```
D   NARPS=(V/S)       NORMAL ATTRITION RATE PER SHELL
D   NAP=(V)        NUMBER ADVANCING FOR PLOTS
D   OARS=(1)        OLD VALUE OF RED ARRIVAL SWITCH
D   OCCS=(1)        OLD VALUE OF COMPANY COLUMN SWITCH
D   OMBF=(V*(K/H))    OLD VALUE OF MOMENTUM IN BATTALION
/*  FORMATION
D   OMCC=(V*(K/H))    OLD VALUE OF MOMENTUM IN COMPANY
/*  COLUMN
D   OSBF=(K/H)       OLD VALUE OF SPEED IN BATTALION
/*  FORMATION
D   OSCC=(K/H)        OLD VALUE OF SPEED IN COMPANY COLUMN
D   PCDR=(V/H)        PLATOON COLUMN DEPLOYMENT RATE
D   PD=(H)        PERCEPTION DELAY
D   PCDR1=(V/H)        PLATOON COLUMN DEPLOYMENT RATE 1
D   PD2=(1)        PROPORTION OF BATTALION FORMATION
/*  DISTANCE AT WHICH BLUE FIRE RESTARTS
D   PDTE=(K)        PERCEIVED DISTANCE TO ENEMY
D   PD4=(1)        PROPORTION OF COMPANY COLUMN DISTANCE
/*     AT WHICH BLUE FIRE RESTARTS
D   PDCCD=(K)        PLANNED DISTANCE TO COMPANY COLUMN
/*  DEPLOYMENT
D   PD3=(1)        PROPORTION OF COMPANY COLUMN DISTANCE
/*     AT WHICH BLUE FIRE CEASES
D   PD1=(1)        PROPORTION OF BATTALION FORMATION
/*     DISTANCE AT WHICH BLUE FIRE CEASES
D   PD22=(1)        PROPORTION OF BATTALION FORMATION
/*     DISTANCE AT WHICH BLUE FIRE FINALLY CEASES
D   PD44=(1)        PROPORTION OF COMPANY COLUMNS DISTANCE
/*     AT WHICH BLUE FIRE FINALLY CEASES
D   PPSA=(V/S)        PRODUCTIVITY PER SHELL ATTRITION
D   PDPCD=(K)        PLANNED DISTANCE TO PLATOON COLUMN
/*     DEPLOYMENT
D   PPS=((K/H)/S)     PRODUCTIVITY PER SHELL (SPEED)
D   PRTPER=(H)        PRINTING INTERVAL
D   PSCC=(K/H)        PLANNED SPEED IN COMPANY COLUMNS
D   PSBF=(K/H)        PLANNED SPEED IN BATTALION FORMATION
D   P1=(1)        COMPANY COLUMN POLICY SWITCH
D   P=(1)        PERCEPTION FACTOR
D   PWDRA=(S/H)        PLANNED WEAPON DELIVERY RATE AIRCRAFT
D   PWDRG=(S/H)        PLANNED WEAPON DELIVERY RATE ARITILERY
D   RCDBF=(K/H)        RATE OF CHANGE OF DISTANCE IN
/*      BATTALION FORMATION
D   RCDCC=(K/H)        RATE OF CHANGE OF DISTANCE IN
```

```
/*      COMPANY COLUMNS
D  RDSBF=((K/H)/H)   RATE OF DECREASE IN SPEED IN
/*      BATTALION FORMATION
D  RDSCC=((K/H)/H)   RATE OF DECREASE IN SPEED IN
/*      COMPANY COLUMNS
D  RISBF=((K/H)/H)   RATE OF INCREASE IN SPEED IN
/*      BATTALION FORMATION
D  RISCC=((K/H)/H)   RATE OF INCREASE IN SPEED IN
/*      COMPANY COLUMNS
D  SPP=(K/H)      SPEED FOR PLOTTING
D  SS=(1)      SPEED CRITERION SWITCH
D  ST=(H)      TABLE FOR SPEED CORRECTION TIMES
D  STT=(H)      SCHEDULED TOTAL TIME
D  SZS=(1)      SPEED / SIZE RATIO SWITCH
D  TCS=(H)      TIME TO CORRECT SPEED IN BATTALION
/*      FORMATION
D  TCSC=(H)      TIME TO CORRECT SPEED IN COMPANY COLUMN
D  TDF=(1)      TOTAL DENSITY FACTOR
D  TD=(K)      TOTAL DISTANCE
D  TH=(1)      THOUSAND
D  TTCCD=(H)      TOTAL TIME TO COMPANY COLUMN DEPLOYMENT
D  TIME=(H)      TOTAL ADVANCING TIME
D  UIAA=(V)      UNITS IN ASSEMBLY AREA
D  UIAAX=(V)      INITIAL UNITS IN ASSEMBLY AREA
D  VDCCD=(K)      VARIABLE DISTANCE TO COMPANY COLUMN
/*      DEPLOYMENT
D  WAT=(1)      TABLE FOR WEAPON ACCURACY MULTIPLIER
D  WAM=(1)      WEAPON ACCURACY MULTIPLIER
D  WDRGS=(S/H)      WEAPON DELIVERY RATE BY ARTILLERY
/*      ON SPEED CRITERION
D  WDRAS=(S/H)      WEAPON DELIVERY RATE BY AIRCRAFT
/*      ON SPEED CRITERION
D  WDAS=(1)      WEAPON DELIVERY BY AIRCRAFT SWITCH
D  WDRG=(S/H)      ACTUAL WEAPON DELIVERY RATE ARTILLERY
D  WDRA=(S/H)      ACTUAL WEAPON DELIVERY RATE AIRCRAFT
D  WDGS=(1)      WEAPON DELIVERY SWITCH FOR ARTILLERY
/*      FIRE
D  WDRAD=(S/H)      WEAPON DELIVERY RATE BY AIRCRAFT ON
/*      DISTANCE CRITERION
D  WDRGD=(S/H)      WEAPON DELIVERY RATE BY ARTILLERY ON
/*      DISTANCE CRITERION
D  WDRAM=(S/H)      WEAPON DELIVERY RATE BY AIRCRAFT ON
/*      MOMENTUM CRITERION
```

```
D   WDRGM=(S/H)       WEAPOM DELIVERY RATE BY ARTILLIARY
/*     ON MOMENTUM CRITERION
D   ZS=(K/H)        ZERO SPEED
D   ZR=((K/H)/H)      ZERO RETARDATION
D   ZD=(K)        ZERO DISTANCE
D   ZAODR=(V/H)       ZERO ATTRITION OR DEPLOYMENT RATE
D   ZFR=(S/H)        ZERO FIRING RATE
D   ZT=(H)        ZERO TIME
D   ZA=(S)        ZERO AMMUNITION
RUN VARIABLE DISTANCE FORMATION CHANGE,HIGH FIRE RATE,
     MOMENTUM CRITERION
```

Table 2   Notes on Equations for the Armoured Advance Model of Chapter 8

1. *Switches*

A number of switches (0,1 variables) are created using CLIP functions. These variables are used throughout the model as multipliers to control a variety of activities as follows:

P1 : when = 1 Red changes formation at a fixed distance
    : when = 0 Red changes formation at a variable distance

Examples of Use : for ease of experimentation
              : in the equation for company column deployment rate to apply different conditions for the activation of company column deployment.

DS, SS, MS : when these variables = 1 Blue delivers fire on a distance, speed or momentum criterion, respectively.

Examples of Use : for ease of experimentation
              : to allow one variable for the weapon delivery rate per aircraft/artillery to be created, which takes on a different value dependent on the value of these switches.

WDAS, WDGS when = 1 weapon delivery is from aircraft/artillery, respectively.

Examples of Use : by multiplying (1-WDAS) by the rate of increase in speed in battalion formation (RISBF), this latter variable can operate and allow Red speed to recover when Blue is not firing.

: by multiplying WDAS by the rate of decrease of speed in battalion formation (RDSBF), this latter variable will operate only when Blue is firing.

: WDGS and (1-WDGS) are used in a similar way to WDAS and (1-WDAS) on rates of speed increase and decrease in company columns.

BFS, CCS : when = 1 there are vehicles advancing in battalion formation/company columns.

Examples of Use : by multiplying BFS by the rate of decrease of speed in battalion formation, this latter variable can only operate when there are vehicles in battalion formation to be slowed down. This also applies to the attrition rate in battalion formation (ARBF).

: CCS is used in a similar way as a condition for the attrition rate and speed reduction on vehicles in company columns.

2. *Calculation of terminal conditions*

The calculation of the time, size and momentum of arrival of the Red force at the Blue position involves isolating the value of these variables at a particular time.

These calculation are achieved by firstly creating an arrival switch (ARS) which changes to 1 when either the cumulative distance of advance = 1 or when the simulation time = the simulation length. Secondly, an old value of this switch is calculated (OARS). That is, the value of the switch 1 DT ago. Hence, the unique event of Red arrival occurs during the DT of the simulation when ARS = 1 and OARS = 0, or when the product of ARS and (1-OARS) = 1.

This product is used in the arrival time (ART) equation which is a level equation fed by a rate of time. ART will be zero until the event of arrival when it will change to the value of time at that point.

The product is also used in the arrival size (ARSZ) equation in a similar way. The arrival momentum is calculated from the arrival time and the arrival size.

The calculation of a company column switch and its old values can also be used to determine the time, size and momentum of Red at the point of deployment to company column formation.

3. *Positive Levels*

Throughout the model all rates which deplete levels are formulated as the MIN of the desired depletion rate, or the MAX of zero or the value of the level divided by 1 DT.

This is to ensure that a rate is never applied to a level which will deplete it in the next DT of the simulation by more than the quantity it contains.

Appendix 5

(X or /* denotes a continuation line)

Table 1 DYSMAP2 Equations for the Curve Fitting Model of Chapter 9

```
* curve fitting
NOTE
NOTE        APPENDIX 5      TABLE 1    (MODEL CURVEB)
NOTE
NOTE   DYSMAP2 EQUATIONS FOR THE CURVE FITTING MODEL
NOTE            (RESULTS IN FIGURES 9.8 AND 9.9)
NOTE
A DEQ.K=CON+SLP*TIME.K-AMPL*SIN(TP*TIME.K/PRD)
C TP=6.283
A DEM.K=TABHL(DTAB,TIME.K,LL,UL,INT)
T DTAB=430/447/440/316/397/375/292/458/400/350/284/
X 400/483/509/500/475/500/600/700/700/725/600/432/615
C LL=0
C UL=23
C INT=1
A DEV.K=DEQ.K-DEM.K
L SSD.K=SSD.J+DT*SQD.J
N SSD=0
A SQD.K=DEV.K*DEV.K
C CON=430
C SLP=2
C AMPL=200
C PRD=50
C DT=1
C LENGTH=23
C PRTPER=1
N TIME=0
NOTE C PLTPER=1
PRINT DEQ,DEM
```

```
NOTE
NOTE      DOCUMENTATION
NOTE
D TP=(1) SINE WAVE CONSTANT
D DEQ=(U/M) DEMAND EQUATION
D DEM=(U/M) DEMAND
D DEV=(U/M) DEVIATION
D LL=(M) LOWER LIMIT OF DTAB
D UL=(M) UPPER LIMIT OF DTAB
D INT=(M)  INTERVAL FOR DTAB
D SQD=(U/M)*(U/M) SQUARED DEVIATION
D SSD=((U*U)/M)      SUM OF SQUARED DEVIATIONS
D DTAB=(U/M)   DEMAND TABLE
D CON=(U/M) CONSTANT
D SLP=(U/M/M)  SLOPE
D AMPL=(U/M) AMPLITUDE
D PRD=(M)       PERIOD
D LENGTH=(M)   SIMULATION LENGTH
D TIME=(M)      SIMULATION TIME
D DT=(M)          SIMULATION INTERVAL
D PRTPER=(M)   SIMULATION PRINT INTERVAL
RUN FIGURE 9.8 BASE MODEL FOR CURVE FITTING
C SLP=3.998
C AMPL=109.475
C PRD=26.060
RUN FIGURE 9.9 CURVE FITTING MODEL WITH OPTIMISED
    PARAMETERS
```

Table 2 DYSMAP2 Equations for the S-Curve Model of Chapter 9

```
* S-CURVE MODEL
NOTE
NOTE      APPENDIX 5     TABLE 2 (MODEL S-CB)
NOTE
NOTE      DYSMAP2 EQUATIONS FOR THE S-CURVE MODEL
NOTE         OF CHAPTER 9
NOTE            (RESULTS IN FIGURES 9.11-9.13)
NOTE
L CNA.K=CNA.J+DT*DSR.JK
N CNA=0
R DSR.KL=TABHL(PTAB,CNA.K,LL,UL,INT)
C LL=0
C UL=400
```

```
C INT=100
T PTAB=3/3/3/3/3
A CNAD.K=TABHL(CTAB,TIME.K,LL1,UL1,INT1)
C LL1=0
C UL1=100
C INT1=25
T CTAB=0/50/150/330/400
A DEV.K=(CNAD.K-CNA.K)*(CNAD.K-CNA.K)
L CDEV.K=CDEV.J+DT*DEV.J
N CDEV=0
PRINT 1)CNA
PRINT 2)CNAD
PRINT 3)DSR
C DT=1
C LENGTH=100
C PRTPER=5
N TIME=0
NOTE
NOTE        DOCUMENTATION
NOTE
D CDEV=(A*A*M)   SUM OF SQUARED DEVIATIONS
D CNA=(A)     CUMULATIVE ADOPTERS
D CNAD=(A)    CUMULATIVE ADOPTERS DESIRED
D DEV=(A*A)    SQUARED DEVIATIONS
D DSR=(A/M)    DIFFUSION RATE
D PTAB=(A/M)    TABLE LINKING DIFFUSION RATE TO
/*  CUMULATIVE ADOPTERS
D CTAB=(A)    TABLE OF CUMULATIVE ADOPTERS DESIRED
D LENGTH=(M)    SIMULATION LENGTH
D PRTPER=(M)    SIMULATION PRINTING INTERVAL
D DT=(M)    SIMULATION INTERVAL
D LL=(M)    LOWER LIMIT FOR TABLE PTAB
D UL=(M)    UPPER LIMIT FOR TABLE PTAB
D INT=(M)    INTERVAL FOR TABLE PTAB
D LL1=(M)    LOWER LIMIT FOR TABLE CTAB
D UL1=(M)    UPPER LIMIT FOR TABLE CTAB
D INT1=(M)    INTERVAL FOR TABLE CTAB
RUN   FIGURE 9.11 CONSTANT DIFFUSION RATE
T PTAB=1.549/4.127/7.309/6.442/0.808
RUN   FIGURE 9.13 OPTIMISED DIFFUSION RATE
```

Appendix 6

DYSMAP2 Functions

The following are a selection of the DYSMAP2 functions used in this book. A full list can be found in the DYSMAP2 User Manual (Dangerfield and Vapenikova, University of Salford 1987).

CLIP function

This function provides a conditional choice of two values for a variable, i.e. CLIP(P,Q,R,S) selects P if R > = S or Q if R < S.

MAX function

This function selects the maximum of two values, i.e. MAX(P,Q) selects P if P > = Q or Q if P < Q.

PULSE function

It is often useful to make sharp changes in a level, for example to re-zero a cumulation after a period of time, or to inject sudden shocks into the system. This is done by the PULSE function which injects a pulse of a given height, lasting for one DT and repeated at intervals. The format is PULSE(HGHT,FRST,INTVL) where HGHT is the pulse height, FRST is the TIME of the first pulse and INTVL is the intervals between pulses.

STEP function

This function generates sudden step changes in a variable. The format is:- STEP(HGHT,STM) where HGHT is the step height STM is the TIME the step occurs.

DELAY Functions

Delays fall into two categories: resource delays and information delays. Resource delays can only occur in the flow of a physical asset, e.g. material (products), orders, money, capital equipment or people. The 'outflow' is then a delayed version of the 'inflow'. The time gap between the inrate and the outrate is the average delay time. This is normally a constant, but it can be a variable.

A simple first-order or exponential delay consists of an inrate, a single level in which the material is held during the delay duration, DEL, and an outrate. A second order delay can be constructed by considering the outrate of the first delay to be the inrate to another first-order delay. The outflow rate of the second delay is the final outflow. The two delays are said to be cascaded. If DEL is the average total time that an entity spends in a second-order delay, each of the two first-order delays is given a delay duration of DEL/2. The order of the delay is the number of internal levels in that delay. A third-order delay will have the internal level variable split into three equal level components each with a delay time DEL/3.

N.B. If DT is too large the delays become unstable because the internal levels acquire or lose too much of the quantity being delayed and this leads to large fluctuations in the output rate. To prevent instability from delays and to ensure that output from the delays in the model approximates reasonably closely to that of the system delays, the value chosen for DT should be one quarter of the smallest first-order delay (one half prevents instability and one tenth gives good accuracy, so one quarter is a good compromise), or one twelfth of the smallest third order delay period specified in the model.

In general, DT should be selected as follows:

DT < = min(DEL/4n)

> where DEL = length of the delay
> n = order of the delay
> min = select the minimum of (DEL/4n) if two or more delays
> are present, in the model.

The delay functions available in DYSMAP2 and used in this book are:

(1) Resource delays

```
IN.KL= DELAY1 (IN.JK,DEL)
IN.KL= DELAY2 (IN.JK,DEL)
IN.KL= DELAY3 (IN.JK,DEL)
```

(2) Information delays

```
IN.KL= DLINF1(IN.JK,DEL)  (synonym SMOOTH)
IN.KL= DLINF2(IN.JK,DEL)
IN.KL= DLINF3(IN.JK,DEL)
```

where IN is the input to the delay and DEL is the average length of the delay, which may be a variable.
Expansions of the first- and third-order delay equations given above can be found in Figures 5.14 and 5.15.

TABLE function

The general format is TABLE(TAB,X,LO,HI,INCR)

where TAB is the name of the table,
X is the independent variable, which may be an expression,
LO is the lowest value in the range of the independent variable,
HI is the highest value in the range of the independent variable and
INCR is the increment of the independent variable

Corresponding to each table function, a list of the values of the dependent variable (i.e. those values read off the vertical axis of the graph) must be specified on a T-statement, with the name of the table (TAB). The number of values on this T-statement is equal to (HI-LO)/INCR + 1.

Note that the independent variable, X, is not permitted to go outside the range specified. If it does, a failure occurs in the running of the model and DYSMAP2 produces a run-time error message identifying the table in question.

Although X is normally a single variable, it can be an arbitrary expression, but it is usually better to define X as an auxiliary variable.

TABHL function

If it is required that the independent variable be permitted to exceed the range of the table, then the TABHL function should be used rather than the TABLE function. This function continues the values given by the first and last values in the table, and uses these for the result if the independent variable is respectively less than or greater than the table range respectively.
The general format is:

TABHL(TAB,X,LO,HI,INCR)

where the arguments have the same meaning as in the TABLE function above. This is the version of the table function used throughout this book.

Index